CONTEMPORARY POLITICAL PHILOSOPHY

CONTEMPORARY POLITICAL PHILOSOPHY

An Introduction

WILL KYMLICKA

CLARENDON PRESS · OXFORD

OXFORD

UNIVERSITY PRESS

Great Clarendon Street, Oxford OX2 6DP

Oxford University Press is a department of the University of Oxford.
It furthers the University's objective of excellence in research, scholarship,
and education by publishing worldwide in

Oxford New York

Athens Auckland Bangkok Bogotá Buenos Aires Calcutta
Cape Town Chennai Dar es Salaam Delhi Florence Hong Kong Istanbul
Karachi Kuala Lumpur Madrid Melbourne Mexico City Mumbai
Nairobi Paris São Paulo Singapore Taipei Tokyo Toronto Warsaw

with associated companies in Berlin Ibadan

Oxford is a registered trade mark of Oxford University Press
in the UK and in certain other countries

Published in the United States
by Oxford University Press Inc., New York

© Will Kymlicka 1990

The moral rights of the author have been asserted
Database right Oxford University Press (maker)

First published 1990
Reprinted in paperback 1992 (twice), 1993, 1995, 1997, 1999
Reprinted in hardback and paperback 1997

British Library Cataloguing in Publication Data
Kymlicka, Will
Contemporary political philosophy: an introduction.
1. Politics. Theories
I. Title   320.0
ISBN 0-19-827724-5
ISBN 0-19-827723-7 (Pbk)

Library of Congress Cataloging in Publication Data
Kymlicka, Will.
Contemporary political philosophy: an introduction/Will Kymlicka.
Includes bibliographical references and index.
1. Political science—History—20th century.   2. Political science—
Philosophy.   I. Title.
JA83.K95   1990   320.5'09'04—dc20   90-39509
ISBN 0-19-827724-5
ISBN 0-19-827723-7 (Pbk.)

Typeset in Sabon

Printed in Great Britain
on acid-free paper by
Biddles Ltd
Guildford and King's Lynn

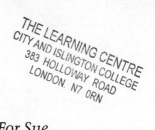

For Sue

Acknowledgements

I would like to thank Richard Arneson, Ian Carter, James Griffin, Sally Haslanger, Brad Hooker, Andrew Kernohan, David Knott, Henry Laycock, Colin Macleod, Susan Moller Okin, Arthur Ripstein, Wayne Sumner, and Peter Vallentyne for helpful comments on parts or all of this book. I owe a special debt to G. A. Cohen, who taught me most of what I know about the aims and methods of political philosophy. He has generously and patiently commented on many of the arguments in this book. My greatest debt is to Susan Donaldson, who has gone over the book with me, line by line, on more than one occasion.

Contents

1

Introduction

1. THE PROJECT

This book is intended to provide an introduction, and critical appraisal, of the major schools of thought which dominate contemporary debates in political philosophy. The material covered is almost entirely comprised of recent works in normative political philosophy and, more particularly, recent theories of a just or free or good society. It does not cover, except incidentally, the major historical figures, nor does it cover many other subjects that were once considered the focal point of political philosophy—e.g. the conceptual analysis of the meaning of power, or sovereignty, or of the nature of law. These were popular topics twenty-five years ago, but the recent emphasis has been on the ideals of justice, freedom, and community which are invoked when evaluating political institutions and policies. I will not, of course, attempt to cover all the recent developments in these areas, but will concentrate on those theories which have attracted a certain allegiance, and which offer a more or less comprehensive vision of the ideals of politics.

One reason for writing this book is my belief that there is a remarkable amount of interesting and important work being done in the field. To put it simply, the intellectual landscape in political philosophy today is quite different from what it was twenty, or even ten, years ago. The arguments being advanced are often genuinely original, not only in developing new variations on old themes (e.g. Nozick's development of Lockean natural rights theory), but also in the development of new perspectives (e.g. feminism). One result of these developments is that the traditional categories within which political theories are discussed and evaluated are increasingly inadequate.

Our traditional picture of the political landscape views political principles as falling somewhere on a single line, stretching from left to right. According to this traditional picture, people on the left believe in equality, and hence endorse some form of socialism, while those on the right believe in freedom, and hence endorse some form of free-market capitalism. In the middle are the liberals, who believe in a wishy-washy mixture of equality and freedom, and hence endorse some form of welfare-state capitalism. There are, of course, many positions in between these three points, and many people accept different parts of different theories. But it is often thought that the best way to understand or describe someone's political principles is to try to locate them somewhere on that line.

There is some truth to this way of thinking about Western political theory. But it is increasingly inadequate. Firstly, it ignores a number of important issues. For example, left and right are distinguished by their views of freedom and justice in the traditionally male-dominated spheres of government and economy. But what about the fairness or freedom of the traditionally female spheres of home and family? Mainstream political theorists from left to right have tended either to neglect these other spheres, or to claim that they do not raise questions of justice and freedom. An adequate theory of sexual equality will involve considerations that simply are not addressed in traditional left–right debates. The traditional picture has also been criticized for ignoring issues of historical context. Theories on both the left and the right seek to provide us with principles we can use to test and criticize our historical traditions and cultural practices. But communitarians believe that evaluating political institutions cannot be a matter of judging them against some independent ahistorical standard. They believe that political judgement is a matter of interpreting the traditions and practices we already find ourselves in. So there are issues of our historical and communal 'embeddedness' which are not addressed in traditional left–right disputes. We cannot begin to understand feminism or communitarianism if we insist on locating them somewhere on a single left–right continuum.

So one problem concerns the narrowness of the traditional picture. This objection is a fairly common one now, and most commentators in the field have tried to bring out the greater range of principles that get invoked in political debate. But there is another feature of the traditional picture which I believe is equally in need of revision. The traditional picture suggests that different theories have different foundational values: the reason that right and left disagree over capitalism is that the left believes in equality while the right believes in freedom. Since they disagree over fundamental values, their differences are not rationally resolvable. The left can argue that if you believe in equality, then you should support socialism; and the right can argue that if you believe in freedom, you should support capitalism. But there is no way to argue for equality over freedom, or freedom over equality, since these are foundational values, with no higher value or premiss that both sides can jointly appeal to. The deeper we probe these political debates, the more intractable they become, for we are left with nothing but conflicting appeals to ultimate, and ultimately opposed, values.

This feature of the traditional picture has remained largely unquestioned, even by those commentators who reject the traditional left–right classifications. Each of the new theories is also assumed to appeal to a different ultimate value. Thus we are told that alongside the older appeal to 'equality' (socialism) and 'liberty' (libertarianism), political theories now appeal to the ultimate values of 'contractual agreement' (Rawls), 'the common good' (communitarianism), 'utility' (utilitarianism), 'rights' (Dworkin), or 'androgyny' (feminism).[1] So we now have an even greater number of ultimate values between which there can be no rational arguments. But this explosion of potential ultimate values raises an obvious problem for the whole project of developing a single comprehensive theory of justice. If there are so many potential ultimate values, why should we continue to think that an adequate political theory can be based on just one of them? Surely the only sensible response to this plurality of proposed ultimate values is to give up the idea of developing a 'monistic' theory of justice. To subordinate all other values to a single overriding one seems almost fanatical.

A successful theory of justice, therefore, will have to accept bits and pieces from most of the existing theories. But if the disagreements between these values really are foundational, how can they be integrated into a single theory? One traditional aim of political philosophy was to find coherent and comprehensive rules for deciding between conflicting political values. But how can we have such comprehensive criteria unless there is some deeper value in terms of which the conflicting values are judged? Without such a deeper value, there could only be *ad hoc* and localized resolutions of conflicts. We would have to accept the inevitable compromises that are required between theories, rather than hope for any one theory to provide comprehensive guidance. And indeed this is what many commentators believe is the fate of contemporary theorizing about justice. Political philosophy is, on this view, drowning in its own success. There has been an explosion of interest in the traditional aim of finding the one true theory of justice, but the result of this explosion has been to make that traditional aim seem wholly implausible.

Is this an accurate picture of the political landscape? Do contemporary political theories appeal to conflicting ultimate values? I want to explore a suggestion, advanced by Ronald Dworkin, that modern political theories do not have different foundational values. On Dworkin's view, every plausible political theory has the same ultimate value, which is equality. They are all 'egalitarian' theories (Dworkin 1977: 179–83; 1983: 24; 1986: 296–301; 1987: 7–8; cf. Nagel 1979: 111). That suggestion is clearly false if by 'egalitarian theory' we mean a theory which supports an equal distribution of income. But there is another, more abstract and more fundamental, idea of equality in political theory—namely, the idea of treating people 'as equals'. There are various ways to express this more basic idea of equality. A theory is egalitarian in this sense if it accepts that the interests of each member of the community matter, and matter equally. Put another way, egalitarian theories require that the government treat its citizens with equal consideration; each citizen is entitled to equal concern and respect. This more basic notion of equality is found in Nozick's libertarianism as much as in Marx's communism. While leftists believe that equality of income or wealth is a precondition for treating people as equals, those on the right believe that equal rights over one's labour and property are a precondition for treating people as equals.

So the abstract idea of equality can be interpreted in various ways, without necessarily favouring equality in any particular area, be it income, wealth, opportunities, or liberties. It is a matter of debate between these theories which specific kind of equality is required by the more abstract idea of treating people as equals. Not every political theory ever invented is egalitarian in this broad sense. But if a theory claimed that some people were not entitled to equal consideration from the government, if it claimed that certain kinds of people just do not matter as much as others, then most people in the modern world would reject that theory immediately. Dworkin's suggestion is that the idea that each person matters equally is at the heart of all plausible political theories.

This is the suggestion I want to explore in this book, for I believe it is as important as any of the particular theories which it attempts to interpret. (One of its advantages is that it makes the quest for a single comprehensive theory of justice seem more intelligible.) Not everyone agrees that each of these theories is based on a principle of equality, and I will be looking at other ways of interpreting them. For example, I will be discussing what it might mean for libertarianism to have freedom as its foundational value, or for utilitarianism to have utility as its foundational value. In each case, I will compare the different interpretations to see which presents the most coherent and attractive account of the theory in question.

If Dworkin's suggestion is correct, then the scepticism many people feel about the possibility of rationally resolving debates between theories of justice may be misplaced or, at any rate, too hasty. If each theory shares the same 'egalitarian plateau'—that is, if each theory is attempting to define the social, economic, and political conditions under which the members of the community are treated as equals—then we might be able to show that one of the theories does a better job living up to the standard that they all recognize. Whereas the traditional view tells us that the fundamental argument in political theory is whether to accept equality as a value, this revised view tells us that the fundamental argument is not whether to accept equality, but how best to interpret it. And that means people would be arguing on the same wavelength, so to speak, even those who do not fit on the traditional left–right continuum. Thus the idea of an egalitarian plateau for political argument is potentially better able to accommodate both the diversity and unity of contemporary political philosophy.

2. A NOTE ON METHOD

It is common in a book of this sort to say something about one's methodology, about how one understands the enterprise of political philosophy, what distinguishes it from other intellectual enterprises, such as moral philosophy, and how one goes about judging its success. I will not say much about these questions here, partly because I do not think there is much that can be said at a general level. Each of the theories examined below answers these questions in a different way—each offers its own account of the division between moral and political philosophy, and its own account of the criteria of successful argument. Evaluating a particular account of the nature of political philosophy, therefore, cannot be separated out from, or done in advance of, evaluating substantive theories of justice.

However, it may be helpful to foreshadow some of the points discussed in later chapters. I believe there is a fundamental continuity between moral and political philosophy, in at least two respects. Firstly, as Robert Nozick puts it, 'moral philosophy sets the background for, and boundaries of, political philosophy. What persons may and may not do to one another limits what they may do through the apparatus of a state, or do to establish such an apparatus. The moral prohibitions it is permissible to enforce are the source of whatever legitimacy the state's fundamental coercive power has' (Nozick 1974: 6). We have moral obligations towards each other, some of which are matters of public responsibility, enforced through public institutions, others of which are matters of personal responsibility, involving rules of personal conduct. Political philosophy focuses on those obligations which justify the use of public institutions. Different theories distinguish public and private responsibility in different ways, but I agree with Nozick that the content of these responsibilities, and the line between them, must be determined by appeal to deeper moral principles.

Secondly, and relatedly, any account of our public responsibilities must fit into a broader moral framework that makes room for, and makes sense of, our private responsibilities. Even where a political theory makes a sharp distinction between public and private responsibility, so that the political principles it endorses have little immediate bearing on rules of personal conduct, it still must not crowd out (in theory or practice) our sense of personal responsibility for helping friends, keeping promises, pursuing projects. This is a problem, I believe, for utilitarian accounts of justice (ch. 2). On the other hand, it is equally true that any account of our personal obligations must make room for what Rawls calls 'the very great values applying to political institutions', such as democracy, equality, and tolerance. For example, it is an important criticism of the 'ethic of care' that it leaves no room for these political values to operate—they are crowded out by the dynamics of ethical caring (ch. 7).

This leaves us with many unanswered questions about the relationship between moral and political philosophy, and about the sorts of convergence and conflict we can expect or tolerate between personal and political values. But these are issues that can only be discussed within the context of particular theories.

As for the criteria by which we judge success in the enterprise of political philosophy, I believe that the ultimate test of a theory of justice is that it cohere with, and help illuminate, our considered convictions of justice. If on reflection we share the intuition that slavery is unjust, then it is a powerful objection to a proposed theory of justice that it supports slavery. Conversely, if a theory of justice matches our considered intuitions, and structures them so as to bring out their internal logic, then we have a powerful argument in favour of that theory. It is of course possible that these intuitions are baseless, and the history of philosophy is full of attempts to defend theories without any appeal to our intuitive sense of right and wrong. But I do not believe there is any other plausible way of proceeding. In any event, the fact is that we have an intuitive sense of right and wrong, and it is natural, indeed unavoidable, that we try to work out its implications—that we seek to do 'what we can to render coherent and to justify our convictions of social justice' (Rawls 1971: 21).

Different theories appeal to our considered convictions in different ways. Utilitarians and libertarians, for example, appeal to them in a more indirect way than liberals or feminists do, and communitarians give our intuitions a quite different status from Marxists. But, again, these are all matters to be discussed in the context of particular theories.

So political philosophy, as I understand it, is a matter of moral argument, and moral argument is a matter of appeal to our considered convictions. In saying this, I am drawing on what I take to be the everyday view of moral and political argument; that is, we all have moral beliefs, these beliefs can be right or wrong, we have reasons for thinking they are either right or wrong, and these reasons and beliefs can be organized into systematic moral principles and theories of justice. A central aim of political philosophy, therefore, is to evaluate competing theories of justice to assess the strength and coherence of their arguments for the rightness of their views.

This will seem a hopeless aim to many people. Some people believe that moral values do not really exist, and hence our 'beliefs' about these values are really just statements of personal preference. As such they cannot be right or wrong, and there is no room for rationally evaluating them. Others believe that while moral beliefs may be right or wrong, there is no way to organize them into systematic principles. Our judgements of justice come from a tacit understanding or sense of appropriateness which tells us how to respond to particular circumstances. Any attempt to formalize these judgements into abstract rules or principles distorts them and produces empty formulas. Still others believe that while we have reasons for our beliefs about justice, and while these reasons may be organized into systematic principles, the only intelligible kinds of reasons and principles are those that appeal to our historical traditions. Justice is a matter of cultural interpretation rather than philosophical argument.

I will consider some of these alternative ways of understanding the enterprise in later chapters. However, I do not believe that these (or other) critiques of the traditional aims of political philosophy are successful. I will not attempt to establish the possibility of rationally defending a comprehensive theory of justice, or refute the various objections to it. In fact, I doubt there is any way to defend that possibility, other than by providing particular arguments for a particular theory. The only way to show that it is possible to advance compelling arguments for the rightness or wrongness of principles of justice is to advance some compelling arguments. The rest of this book is, therefore, the only argument I have for the usefulness of my methodological assumptions. Whether it is a good argument or not is for the reader to decide.

NOTES

1. Versions of this list of 'ultimate values' can be found, with minor variations, in most recent surveys of theories of justice (e.g. Brown 1986; Pettit 1980; Sterba 1988; Campbell 1988; Miller 1976).

2

Utilitarianism

It is generally accepted that the recent rebirth of normative political philosophy began with the publication of John Rawls's *A Theory of Justice* in 1971, and his theory would be a natural place to begin a survey of contemporary theories of justice. His theory dominates contemporary debates, not because everyone accepts it, but because alternative views are often presented as responses to it. But just as these alternative views are best understood in terms of their relationship to Rawls, so understanding Rawls requires understanding the theory to which he was responding—namely, utilitarianism. Rawls believes, rightly I think, that in our society utilitarianism operates as a kind of tacit background against which other theories have to assert and defend themselves. So that is where I too will begin.

Utilitarianism, in its simplest formulation, claims that the morally right act or policy is that which produces the greatest happiness for the members of society. While this is sometimes offered as a comprehensive moral theory, I will focus on utilitarianism as a specifically political morality. On this view, utilitarian principles apply to what Rawls calls 'the basic structure' of society, not to the personal conduct of individuals. However, since much of the attraction of utilitarianism as a political morality stems from the belief that it is the only coherent and systematic moral philosophy, I will briefly discuss some features of comprehensive utilitarianism in section 3. In either its narrow or comprehensive version, utilitarianism has both devoted adherents and fierce opponents. Those who reject it say that the flaws of utilitarianism are so numerous that it cannot help but disappear from the landscape (e.g. Williams 1973). But there are others who find it hard to understand what else morality could be about than maximizing human happiness (e.g. Hare 1984).

1. TWO ATTRACTIONS

I will start with utilitarianism's attractions. There are two features of utilitarianism that make it an attractive theory of political morality. Firstly, the goal which utilitarians seek to promote does not depend on the existence of God, or a soul, or any other dubious metaphysical entity. Some moral theories say that what matters is the condition of one's soul, or that one should live according to God's Divine Will, or that one's life goes best by having everlasting life in another realm of being. Many people have thought that morality is incoherent without these religious notions. Without God, all we are left with is a set of rules—'do this', 'don't do that'—which lack any point or purpose.

It is not clear why anyone would think this of utilitarianism. The good it seeks to promote—happiness, or welfare, or well-being—is something that we all pursue in our own lives, and in the lives of those we love. Utilitarians just demand that the pursuit of human welfare or utility (I will be using these terms interchangeably) be done impartially, for everyone in society. Whether or not we are God's children, or have a soul, or free will, we can suffer or be happy, we can all be better or worse off. No matter how secular we are, we cannot deny that happiness is valuable, since it is something we value in our own lives.

A distinct but related attraction is utilitarianism's 'consequentialism'. I will discuss what exactly that means later on, but for the moment its importance is that it requires that we check to see whether the act or policy in question actually does some identifiable good or not. We have all had to deal with people who say that something—homosexuality, for example (or gambling, dancing, drinking, swearing, etc.)—is morally wrong, and yet are incapable of pointing to any bad consequences that arise from it. Consequentialism prohibits such apparently arbitrary moral prohibitions. It demands of anyone who condemns something as morally wrong that they show *who is wronged*, i.e. they must show how someone's life is made worse off. Likewise, consequentialism says that something is morally good only if it makes someone's life better off. Many other moral theories, even those motivated by a concern for human welfare, seem to consist in a set of rules to be followed, whatever the consequences. But utilitarianism is not just another set of rules, another set of 'do's' and 'don'ts'. Utilitarianism provides a test to ensure that such rules serve some useful function.

Consequentialism is also attractive because it conforms to our intuitions about the difference between morality and other spheres. If someone calls certain kinds of consensual sexual activity morally wrong because they are 'improper', and yet cannot point to anyone who suffers from them, then we might respond that the idea of 'proper' behaviour being employed is not a moral one. Such claims about proper behaviour are more like aesthetic claims, or an appeal to etiquette or convention. Someone might say that punk rock is 'improper', not legitimate music at all. But that would be an aesthetic criticism, not a moral one. To say that homosexual sex is 'improper', without being able to point to any bad consequences, is like saying that Bob Dylan sings improperly—it may be true, but it is not a moral criticism. There are standards of propriety that are not consequentialist, but we think that morality is more important than mere etiquette, and consequentialism helps account for that difference.

Consequentialism also seems to provide a straightforward method for resolving moral questions. Finding the morally right answer becomes a matter of measuring changes in human welfare, not of consulting spiritual leaders, or relying on obscure traditions. Utilitarianism, historically, was therefore quite progressive. It demanded that customs and authorities which had oppressed people for centuries be tested against the standard of human improvement ('man is the measure of all things'). At its best, utilitarianism is a strong weapon against prejudice and superstition, providing a standard and a procedure that challenge those who claim authority over us in the name of morality.

Utilitarianism's two attractions, then, are that it conforms to our intuition that human well-being matters, and to our intuition that moral rules must be tested for their consequences on human well-being. And if we accept those two points then utilitarianism seems to follow almost inevitably. If human welfare is the good which morality is concerned with, then surely the morally best act is the one which maximizes human welfare, giving equal weight to each person's welfare. Those who believe that utilitarianism has to be true are convinced that any theory which denies either of these two intuitions must be false.

I agree with the two core intuitions. If there is a way to challenge utilitarianism, it will not take the form of denying these intuitions. A successful challenge will have to show that some other theory does a better job of spelling them out. I will argue later that there are other theories which do just this. But first we need a closer look at what utilitarianism amounts to. Utilitarianism can be broken down into two parts:

1. an account of human welfare, or 'utility', and
2. an instruction to maximize utility, so defined, giving equal weight to each person's utility.

It is the second claim which is the distinctive feature of utilitarianism, and it can be combined with various answers to the first question. So our final judgement of utilitarianism will depend on our evaluation of the second claim. But it is necessary to begin by considering various answers to the first question.

2. DEFINING UTILITY

How should we define human welfare or utility? Utilitarians have traditionally defined utility in terms of happiness—hence the common but misleading slogan 'the greatest happiness of the greatest number'.[1] But not every utilitarian has accepted such a 'hedonistic' account of human welfare. In fact, there are at least four identifiable positions taken on this question.

(a) Welfare hedonism

The first view, and perhaps the most influential in the utilitarian tradition, is the view that the experience or sensation of pleasure is the chief human good. It is the one good which is an end in itself, to which all other goods are means. Bentham, one of the founders of utilitarianism, said, in a famous quote, that 'pushpin is as good as poetry' if it gives the same intensity and duration of pleasure. If we prefer poetry to pushpin, if we think it a more valuable thing to do with our time, it must be because it gives us more pleasure.

This is a dubious account of why we prefer some activities over others. It is a cliché, but perhaps a true one, that poets often find writing to be painful and frustrating, yet they think it is valuable.

This goes for reading poetry as well—we often find poetry disturbing rather than pleasurable. Bentham might respond that the writer's happiness, like the masochist's, lies precisely in these apparently unpleasant sensations. Perhaps the poet really finds pleasure in being tortured and frustrated.

I doubt it. But we do not have to settle that question, for Robert Nozick has developed an even stronger argument against welfare hedonism (Nozick 1974: 42–5; cf. Smart 1973: 18–21). He asks us to imagine that neuropsychologists can hook us up to a machine which injects drugs into us. These drugs create the most pleasurable conscious states imaginable. Now if pleasure were our greatest good, then we would all volunteer to be hooked for life to this machine, perpetually drugged, feeling nothing but happiness. But surely very few people would volunteer. Far from being the best life we can lead, it hardly counts as leading a life at all. Far from being the life most worth leading, many people would say that it is a wasted life, devoid of value.

In fact, some people would prefer to be dead than to have that sort of life. Many people in the United States sign 'living wills' which demand that they be taken off life support systems if there is no hope of recovery, even if those systems can remove pain and induce pleasure. Whether or not we would be better off dead, we would surely be better off undrugged, doing the things we think worth doing in life. And while we hope we will be happy in doing them, we would not give them up, even for guaranteed happiness.

(b) Non-hedonistic mental-state utility

The hedonistic account of utility is wrong, for the things worth doing and having in life are not all reducible to one mental state like happiness. One response is to say that many different kinds of experiences are valuable, and that we should promote the entire range of valuable mental states. Utilitarians who adopt this account accept that the experience of writing poetry, the mental state accompanying it, can be rewarding without being pleasurable. Utilitarianism is concerned with all valuable experiences, whatever form they take.

But this does not avoid Nozick's objection. Nozick's invention is in fact called an 'experience machine', and the drugs can produce any mental state desired—the ecstasy of love, the sense of accomplishment from writing poetry, the sense of peace from religious contemplation, etc. Any of these experiences can be duplicated by the machine. Would we now volunteer to get hooked up? The answer is still, surely, no.

What we want in life is something more than, or other than, the acquisition of any kind of mental state, any kind of 'inner glow', enjoyable or otherwise. We do not just want the experience of writing poetry, we want to write poetry; we do not just want the experience of falling in love, we want to fall in love; we do not just want the feeling of accomplishing something, we want to accomplish something. It is true that when we fall in love, or accomplish something, we also want to experience it. And we hope that some of those experiences will be happy. But we would not give up the opportunity to fall in love, or accomplish something, even for the guaranteed experience of those things inside an experience machine (Lomasky 1987: 231–3; Larmore 1987: 48–9; Griffin 1986: 13–23).

It is true that we sometimes just want certain experiences. That is one reason people take drugs. But our activities while undrugged are not just poor substitutes for getting what drugs can give us directly. No one would accept that mental states are all that matter, such that being hooked up to an experience machine would be the fulfilment of their every goal in life.

(c) Preference satisfaction

Human well-being is something more than, or other than, getting the right sequence of mental states. A third option is the 'preference-satisfaction' account of utility. On this view, increasing people's utility means satisfying their preferences, whatever they are. People may want to experience writing poetry, a preference which can be satisfied in the experience machine. But they may also want to write poetry, and so forgo the machine. Utilitarians who adopt this account tell us to satisfy all kinds of preferences equally, for they equate welfare with the satisfaction of preferences.

However, if the first two views leave too much out of their account of well-being, this third view leaves too much in. Satisfying our preferences does not always contribute to our well-being. Suppose that we are ordering food for lunch, but some of us want pizza, while others want Chinese food. If the way to satisfy the most preferences is to order pizza, then this sort of utilitarianism tells us to order it. But what if, unbeknownst to us, the pizza we ordered is poisoned, or just rancid? Ordering it now would not promote our welfare. What is good for us can be different from the preferences we currently have. Marxists emphasize this in their theory of false consciousness—e.g. workers have been socialized in such a way as to be unable to see their real interest in socialism. But the same problem arises in less dramatic or controversial ways. We can just lack adequate information, as in the pizza example, or have made mistakes in calculating the costs and benefits of a particular action.

Preferences, therefore, do not define our good. It is more accurate to say that our preferences are predictions about our good. We want to have those things which are worth having, and our current preferences reflect our current beliefs about what those worthwhile things are. But it is not always easy to tell what is worth having, and we could be wrong in our beliefs. We might act on a preference about what to buy or do, and then come to realize that it was not worth it. We often make these sort of mistakes, both in specific decisions, like what food to order, and in 'global preferences' about what sort of life to lead. Someone who has planned for years to be a lawyer may get to law school and realize that they have made a mistake. Perhaps they had a romantic view of the profession, ignoring the competitiveness and drudgery involved. Someone who had planned to remain in their home town may come to realize that it is a parochial way to live, narrow and unchallenging. Such people may regret the years they spent preparing for a certain way of life, or leading that life. They regret what they have done, because people want to have or do the things which are *worth* having or doing, and this may be different from what they *currently prefer* to have or do. The first is what matters to us, not the second (Dworkin 1983: 24–30).

Utilitarianism of the preference-satisfaction variety says that something is made valuable by the fact that lots of people desire it. But that is wrong, and indeed backwards. Having the preference does not make it valuable—on the contrary, its being valuable is a good reason for preferring it. And if it is not valuable, then satisfying my mistaken preference for it will not contribute to my well-being.

My utility is increased, then, not by satisfying whatever preferences I have, but by satisfying those preferences which are not based on mistaken beliefs.

(a) Informed preferences

The fourth account of utility tries to accommodate the problem of mistaken preferences by defining welfare as the satisfaction of 'rational' or 'informed' preferences. Utilitarianism, on this view, aims at satisfying those preferences which are based on full information and correct judgements, while rejecting those which are mistaken and irrational. We seek to provide those things which people have good reason to prefer, that really make their life better off.

This fourth account seems right—the chief human good is the satisfaction of rational preferences.[2] But while this view is unobjectionable, it is extremely vague. It puts no constraints on what might count as 'utility'. Happiness at least had the merit of being in principle measurable. We all have a rough idea of what would increase happiness, what would increase the ratio of pleasurable to painful sensations. A pleasure machine would do that best. But once we view utility in terms of satisfying informed preferences, we have little guidance. For one thing, there are many different kinds of informed preferences, with no obvious way to aggregate them. How do we know whether to promote love, poetry, or pushpin if there is no single overarching value like happiness to measure them by? Moreover, how do we know what preferences people would have if they were informed and rational? For example, philosophers debate whether we should put less weight on desires we will have in the future. Is it irrational to care more about what happens to me today than about what will happen to me tomorrow? The issues involved are complex, yet we need an answer in order to begin the utilitarian calculations.

More puzzling yet is the fact that we have dropped the 'experience requirement'—i.e. informed preferences can be satisfied, and hence our utility increased on this fourth account, without it ever affecting our conscious experiences. Richard Hare, for example, argues that my life goes worse if my spouse commits adultery, even if I never come to know of it. My life is made worse because something that I wanted not to happen has happened. This is a perfectly rational and informed preference, yet my conscious experience may not change when it is satisfied or left unsatisfied (Hare 1971: 131).

I agree with Hare that this should count in determining well-being, that it really does make my life worse. For example, if I continue to act towards my spouse on the belief that she has not committed adultery, then I am now acting on a falsehood. I am living a lie, and we do not want to live such a life (Raz 1986: 300–1). We often say of others that what they do not know will not hurt them. But it is hard to think that way of our own good. I do not want to go on thinking I am a good philosopher if I am not, or that I have a loving family if I do not. Someone who keeps the truth from me may spare me some uncomfortable conscious experiences, but the cost may be to undermine the whole point of my activities. I do philosophy because I think I do it well. If I am not doing it well, then I would rather do something else. I do not want to continue on the mistaken belief that I am doing it well, for I would be wasting my time, and living a lie, which are not things I want to do. If I were to discover that my belief is false, then my activity would have lost its point. And it would have lost its point, not when I discovered that the belief was not true, but when it ceased to be true. At that point, my life became worse off, for at that point I could no longer achieve the goals I was concerned to pursue.

We must accept the possibility that our lives can go worse even when our conscious experiences are unaffected. But this leads to some strange results. For example, Hare extends the notion of utility to include the preferences of dead people. I may have a rational preference that my reputation not be libelled when I am dead, or that my body not be left to rot. It seems bizarre to include the preferences of dead people in utility calculations, but what distinguishes them from the preference that one's spouse do not commit unknown adultery? In both cases, we have rational preferences for things which do not affect our conscious states. I think we can draw some lines here. Not every action which goes against a dead person's preferences makes their life worse off.[3] But there are problems here for the utilitarian. Given these difficulties in determining which preferences increase welfare when satisfied, and the difficulties in measuring welfare even when we do know which preferences are rational, we may find ourselves in a situation where it is impossible to know which act maximizes utility. It may be impossible to know what, from a utilitarian point of view, is the right act. Some people have concluded from this that utilitarianism must be rejected. If we accept the fourth view of welfare, and if welfare cannot be aggregated on that view, then there is no way to know which act maximizes welfare, and we need some other account of the morally right act.

But that is a *non sequitur*. From the fact that we cannot know which act maximizes utility, it does not follow that the morally right act is not the one which maximizes utility. It may just mean that we cannot know which act is morally right. There is no reason to exclude the possibility that humans may not always be able to determine the morally right act. Even if there is an inherent incommensurability of different kinds of value, such that we cannot say that one of a range of value-increasing acts maximizes value, we can still make some less fine-grained rankings, and so make judgements about better or worse acts (Griffin 1986: 75–92).

So utilitarianism, despite its traditional ties to welfare hedonism, is compatible with any of the four accounts of utility. Of course, utilitarianism loses one of its attractions when it leaves hedonism behind. Once we reject the simple accounts of welfare as happiness or preference satisfaction, there is no straightforward method for measuring utility. Utilitarianism does not provide a uniquely simple criterion or scientific method to determine what is right and wrong. But while utilitarianism has no advantage over other theories in measuring human welfare, neither is it disadvantaged. Every plausible political theory has to confront these difficult questions about the proper account of human welfare, and nothing prevents utilitarianism from adopting whatever account its critics favour.[4] If we are to reject utilitarianism, then, it will have to be because of the second part of the theory—i.e. the instruction that we should maximize utility, whichever definition of utility we finally adopt.

3. MAXIMIZING UTILITY

Assuming that we have agreed on an account of utility, should we accept the utilitarian commitment to maximizing utility? Is this the best interpretation of our intuitive commitment to 'consequentialism'? Consequentialism tells us to be concerned with promoting people's utility, and, ideally, we would satisfy all the informed preferences of all people. Unfortunately, that is impossible. There are limited resources available to satisfy people's preferences. Moreover, people's preferences may conflict. So whose preferences should we satisfy? Consequentialism tells us to be concerned with consequences for human welfare, but what if the promotion of one person's welfare conflicts with that of another? Consequentialism needs to be spelled out if we are to answer that question.

How does utilitarianism spell out the idea that we should promote people's utility? Utilitarians say that the right action is the one that maximizes utility—e.g. that satisfies as many informed preferences as possible. Some people's preferences will go unsatisfied, if their preferences conflict with what maximizes utility overall. That is unfortunate. But since the winners necessarily outnumber the losers, there is no reason why the preferences of the losers should take precedence over the more numerous (or more intense) preferences of the winners. For the utilitarian, equal amounts of utility matter equally, regardless of whose utility it is. No one stands in a privileged position in the calculations, no one has a greater claim to benefit from an act than any other. Hence we should bring about consequences which satisfy the greatest number of (informed) preferences amongst people in the society. (This, of course, is the barest sketch of the utilitarian account of consequentialism—I discuss two ways to flesh it out in the next section.)

This commitment to examining the consequences for human well-being is one of the attractions of utilitarianism, as compared to theories which say that we should follow tradition or divine law regardless of the human consequences. But the particular kind of consequentialism in utilitarianism is, I think, unattractive. Where it is impossible to satisfy all preferences, our intuitions do not tell us that equal amounts of utility should always have the same weight. Utilitarianism provides an over-simplified account of our commitment to consequentialism.

Before exploring these issues, however, there are some important differences within utilitarianism that need to be laid out. I have just said that, as utilitarians, we should seek to satisfy the greatest number of preferences. But as I mentioned earlier, there are two different accounts within utilitarianism of who the relevant 'we' is—on one view, all of us are obliged to act according to utilitarian principles, even in our personal conduct (comprehensive moral utilitarianism); on the other view, it is the major social institutions which are specifically obliged to act according to utilitarian principles (political utilitarianism). There are also two different accounts of what it means to 'act according to utilitarian principles'. On one view, this means that the agent should decide how to act by consciously making utilitarian calculations, by trying to assess how different actions would affect the satisfaction of informed preferences (direct utilitarianism); on the other view, the idea of maximizing utility enters only indirectly (if at all) into the agent's decision-making. Morally right actions are those that maximize utility, but agents are more likely to maximize utility by following non-utilitarian rules or habits, than by following utilitarian reasoning (indirect utilitarianism).

These two distinctions can be combined to generate different versions of utilitarianism. Utilitarian principles can be applied more or less comprehensively, and more or less directly. Much of the recent work on utilitarianism has been concerned with exploring these variations, and it seems clear that each version will generate different results. However, I believe that all versions share the same fundamental flaw. I will argue that there is something inherently unattractive about the utilitarian commitment to maximizing utility, and that this flaw is not substantially affected by how (directly or indirectly) or where (comprehensively or to politics) that commitment is applied.[5]

I will begin by considering some problems with utilitarianism as a comprehensive decision-procedure. If we view utilitarianism in this way, then the morally responsible agent will be what David Brink calls a 'U-agent'—someone who decides how to spend her time and resources by calculating the effects on overall utility of the various actions available to her (Brink 1986: 425). This sort of utilitarianism has few contemporary defenders, and many utilitarians would agree with the criticisms I am about to make. But I start with utilitarianism as a comprehensive decision-procedure because it raises in a particularly clear form problems that are also present in the more indirect and political versions of utilitarianism (section 5). Moreover, the issues raised in this section, concerning the proper scope of personal relationships, will reappear in later chapters.

Imagine then that we are U-agents, and that we can calculate which act produces the most utility.[6] Should we base our actions on these utilitarian calculations? There are two main objections to utilitarian decision-making—it excludes the special obligations we have to particular people, and it includes preferences which should not be counted. These two problems stem from the same basic flaw, but I will examine them separately.

(a) Special relationships

U-agents who base their actions on utilitarian calculations assume that each person stands in the same moral relationship to them. But this does not allow for the possibility that I could have special moral relationships to my friends, family, lenders, etc., that I could be under a greater obligation to them than to other possible beneficiaries of my actions. Our intuitions tell us that there are such special obligations, and that they should be fulfilled even if those to whom I am not especially obligated would benefit more.

Consider a loan. It is part of our everyday morality that people come to have differential entitlements in virtue of having loaned money in the past. If someone lends me $10, then she is entitled to receive $10 back from me, even if someone else could make better use of the money. Utilitarian reasoning disregards such backward-looking entitlements, for it says that only forward-looking consequences matter. For the U-agent, the moral value of an act lies solely in its causal properties of producing desirable states of affairs. Hence what I ought to do is pull on the causal lever which will produce the maximal amount of utility for the system as a whole. In deciding how to spend my $10, I must look at all the potential preference satisfactions of people (including myself) and determine which action will maximize them. It is of no interest to the U-agent, in and of itself, that one of those people loaned me the $10, or that someone else performed some service for me on the understanding that she would receive the money. It may be that if the utilities work out in a certain way, I ought to repay the loan, or fulfil my contract. But the process of deciding what to do will go on in exactly the same way as if I had not borrowed or promised the money.

This is counter-intuitive, for most of us would say that the 'past circumstances or actions of people can create differential entitlements or differential deserts to things' (Nozick 1974: 155). The person who lent me $10 has, by that very act, acquired an entitlement to the $10 I am now considering spending, even if some other use of the money would maximize happiness. Does this conflict with our view that morality should be about consequences for human welfare? No, for in saying that I should repay the loan, I am simply saying that I have a greater obligation, at this point in time, to promote my lender's welfare than to help others. We should repay the loan, not because we do not care about the harms and benefits which arise from that act, but because one benefit in particular has special weight.

Unlike the hard-line non-consequentialist, we need not say that these entitlements are indefeasible by any calculation of overall social consequences. If repaying the loan would somehow lead to nuclear destruction, then we clearly ought not to repay the loan. But we can say that there is a duty to repay loans and fulfil contracts which has some independent weight, to be considered alongside the moral weight of overall social benefits. The existence of past entitlements on the part of particular people partially pre-empts, or constrains, the utilitarian quest to maximize the general good. Averting a disastrous drop in welfare is a good reason for using the money in a different way, but the mere fact that repaying the loan does not maximally increase welfare is not a good reason. Not to repay the loan just because it does not maximally increase utility is to ignore the special nature of our obligation to the lender.

This is so firmly entrenched in our moral consciousness that many utilitarians have tried to give a utilitarian account of the weight we attach to promises. They point out the many by-products of breaking a promise. For example, while someone other than the lender may be able to make better use of the money, the lender will feel resentment at being deprived of a promised benefit, a disutility so great that it outweighs the increased utility achieved by giving the money to someone else (Hare 1971: 134). But this gets things backwards. We do not feel that breaking promises is wrong because it produces feelings of resentment. Rather, cheating on promises produces feelings of resentment because it is wrong (cf. Williams 1973: 143). Another utilitarian tactic is to point out that promises create expectations which people depend on. Moreover, failing to repay the loan will jeopardize the lender's willingness to lend in the future, and thereby jeopardize a valuable social institution. So utilitarians respond by pointing out that repaying loans is more likely to maximize utility than one might initially think (Sartorius 1969: 79–80).

This may be true, but it does not solve the problem. It still implies, for example, that 'if you have employed a boy to mow your lawn and he has finished the job and asks for his pay, you should pay him what you promised only if you cannot find a better use for your money' (Sartorius 1969: 79). The U-agent's reasoning, while more complex than one might initially think, still fails to recognize any special relationship between employer and employee, or lender and borrower. Some utilitarians are prepared to accept this result. Rolf Sartorius, for example, says that if the usual factors do not ensure that payment maximizes utility, i.e. if the boy 'is not likely to publicize my breaking my promise to him too loudly, appears to have a reservoir of trust in mankind generally, and any sum I could give him really would do more good if contributed to UNICEF, then the conclusion on act-utilitarian grounds must be that I should give the money to UNICEF. But is this really absurd?' (Sartorius 1969: 80). Yes, this is absurd. What is absurd here is not necessarily the conclusion but the fact that the boy's having actually performed the job, or that I had actually promised him the money, never enters into the decision as such. Notice that the consequences Sartorius mentions would be exactly the same even if the boy had not actually mowed the lawn, but simply (falsely) believed that he had done so, or falsely believed that I had promised him the money. The fact that the boy actually mowed the lawn, or that I had promised him the money, does not matter to the U-agent because nothing we could do or say could ever put us in a special moral relationship such that my obligation to him is greater than my obligation to others. No matter what the boy has done or I have said, he can never have a greater claim on my actions than anyone else.

In our everyday view, the existence of a promise creates a special obligation between two people. The U-agent, however, treats promises and contracts, not as creating special moral ties to one person, but as simply adding new factors into the calculation of overall utility. The everyday view says that I should repay loans *regardless* of whether it maximizes utility. The U-agent says that I should repay the loan *because* it maximizes utility. The boy has no greater claim on me than others, he just is likely to benefit more than they are, and so repayment is the best way to fulfil my utilitarian obligation.

But that is not what a promise is—'To make a promise is not merely to adapt an ingenious device for promoting the general well-being; it is to put oneself in a new relation to one person in particular, a relation which creates a specifically new *prima facie* duty to him, not reducible to the duty of promoting the general well-being of society' (Ross 1930: 38). For U-agents, everyone (including oneself) stands in exactly the same moral position—i.e. everyone is an equally deserving possible beneficiary of one's actions. But this is too flat a picture of the moral landscape, for some people 'may also stand to [one] in the relation of promisee to promiser, of creditor to debtor, of wife to husband, of child to parent, of friend to friend, of fellow countryman to fellow countryman, and the like; and each of these relations is the foundation of a *prima facie* duty' (Ross 1930: 19).

The problem here goes deeper than an inadequate account of promises. The U-agent cannot accommodate the importance of any of our commitments. We all have commitments—to family, political causes, work—which form the focal point of our lives and give some identity to our existence. But if I am to act as a U-agent, then, in each of my decisions, my commitments must be simply added in with all the projects of other people, and be sacrificed when I can produce more utility by promoting someone else's projects. That may sound admirably unselfish. But it is in fact absurd. For it is impossible to be genuinely committed to something and yet be willing to sacrifice that commitment whenever something else happens to maximize utility. Utilitarian decision-making asks that I consider my projects and attachments as no more worthy of my help than anyone else's. It asks, in effect, that I be no more attached to my commitments than to other people's. But that is no different from saying that I should not really be attached to my projects at all. As Bernard Williams puts it,

If you are a person who whole-heartedly and genuinely possesses some of these admirable [projects, affections, and commitments], you cannot also be someone in whose thought and action the requirements of utilitarianism are unfailingly mirrored, nor could you wish to be such a person.... Utilitarianism must reject or hopelessly dilute the value of these other dispositions, regressing to that picture of man which early utilitarianism frankly offered, in which he has, ideally, only private or otherwise sacrificeable projects, together with the one moral disposition of utilitarian benevolence. (Williams 1981: 51, 53)

Now it is true and important that I should respect the legitimate commitments of others. But the way to do this is not to consider them as having an equal claim on my time and energy to that of my own projects. Such an attitude is psychologically impossible, and undesirable even if possible. A valuable human life, on just about anyone's account of it, is one filled with attachments that structure one's life, that give some direction to it. It is the prospect of subsequent achievement or progress in such a commitment that makes our current actions meaningful. As a U-agent, however, one's actions will be determined almost wholly independently of one's commitments. The U-agent's decisions will be 'a function of all the satisfactions which he can affect from where he is: and this means that the projects of others, to an indeterminately great extent, determine his decision' (Williams 1973: 115). The U-agent will have few choices about how to lead his life, few opportunities to act on considerations of the kind of person he is, or wants to become. He will thus have little room for the things we associate with the very idea of 'leading a life'. These will all be submerged beneath the question of which causal levers are optimific.

If I am to lead my own life, there must be room in which I am free to form my own commitments, including the sorts of contracts and promises discussed above. The problem of not allowing people to create special obligations to others through promises is just one aspect of the broader problem of not allowing people to set and pursue their own goals. The problem in all of these cases is the U-agent's assumption that each person has an equal claim to benefit from all of his actions.

Does our intuition in favour of meaningful commitments violate the idea that morality concerns consequences? No, for our intuitive commitment to the general idea of consequentialism never included a commitment to the continuous impartial determination of our actions by the preferences of others, to the exclusion of special relationships and projects. This is simply too crude an interpretation of our belief in consequentialism.

(b) Illegitimate preferences

A second problem with utilitarianism as a decision-procedure concerns its demand, not that each person be given equal weight in our decision-making, but that each source of utility (e.g. each kind of preference) be given equal weight. Consider racial discrimination in a mainly white society. A government health care policy might plan to build one hospital for every 100,000 people, regardless of their race. But a number of whites prefer that blacks do not have equal health care, and when the utility calculations are done, it turns out that utility is maximized by depriving blacks of an equal share of health care (or school facilities etc.). Or what if the very sight of known homosexuals deeply offends the heterosexual majority? Perhaps utility is maximized if openly homosexual people are publicly punished and thrown in jail. Or what about an alcoholic on skid row who has no friends, is offensive to many people, and a nuisance to everyone, begging for money and cluttering up public parks? Perhaps utility would be maximized if we quietly took such people and killed them, so they would not be seen, and would not be a drain on social resources in jail. Some of these preferences are of course uninformed, and so satisfying them would not actually yield any utility (assuming we have abandoned the crude hedonistic accounts of utility). But the desire to deny the rights of others is not always uninformed, and even on the best account of utility, the satisfaction of these preferences can be a genuine source of utility for some people. As Rawls puts it, such preferences are 'unreasonable', from the point of view of justice, but are not necessarily 'irrational', from the point of view of an individual's utility (Rawls 1980: 528–30). If this sort of utility is counted, it may lead to discrimination against unpopular minorities.

Our everyday morality tells us that such preferences are unfair, and should not be counted. That racists want a group of people mistreated is no reason at all to give that group less health care. The racists' desire is illegitimate, so whatever utility would come from satisfying that preference has no moral weight. Even where there is no direct prejudice, there may be unfair preferences which should not count. Someone may wish that blacks do not move into their neighbourhood, not because they actively dislike blacks—they may not care one way or the other—but because others dislike blacks, and so the property value of their home will decrease. Such a preference that blacks be excluded from a neighbourhood is not prejudiced in the same way a racist's is. But it is still an illegitimate preference, since it requires that something be wrongfully taken from blacks. In all these cases, utility is maximized by discriminatory treatment, but only as a result of preferences for benefits which are wrongfully taken from others. Preferences like that, preferences for what rightfully belongs to others, have little or no weight in our everyday moral view.

Utilitarians do not accept the claim that preferences for what 'rightfully' belongs to others are illegitimate. For the U-agent there is no standard of what 'rightfully' belongs to anyone prior to the calculation of utility. What is rightfully mine is whatever distribution maximizes utility, so utility-maximizing acts by definition cannot deprive me of my rightful share. But this violates an important component of our everyday morality. Our commitment to the idea of consequentialism does not include a commitment to the idea that each source of utility should have moral weight, that each kind of preference must be counted.

It seems, then, that the U-agent, in trying to maximize utility, is violating, rather than spelling out, our intuitive idea of consequentialism. Some people deny that utilitarian decision-making has these counter-intuitive results. They admit that utilitarian reasoning seems to allow, or even require, acts which violate special relationships or basic rights, whenever such acts would maximize utility. But they claim that these acts would be disallowed if we shifted to a more sophisticated form of utilitarian decision-making. I have been assuming that U-agents apply the test of utility-maximization to particular acts. But 'rule-utilitarians' argue that we should apply the test of utility to rules, and then perform whichever act is endorsed by the best rules, even if another act might produce more utility. Social co-operation requires rule-following, so we should assess the consequences, not simply of acting in a particular way on this occasion, but of making it a rule that we act in that way.

The issue for U-agents, then, is to determine which set of rules is utility-maximizing. Are we better off in utilitarian terms following a rule that instructs us to keep promises, maintain special relationships, and respect rights, or following a rule that subordinates these principles to calculations of utility? The latter, utilitarians argue, would decrease utility. It would make social co-operation difficult, and cheapen the value of human life and liberty. Moreover, people are likely to abuse the power to break promises or discriminate in the name of the public good. Everyone is worse off if we adopt a rule to break promises or discriminate against unpopular groups whenever we think it would maximize utility (Harsanyi 1985).

Some commentators argue that rule-utilitarianism collapses into act-utilitarianism, since we can describe rules in such a detailed and

narrow way as to make them equivalent to acts (Lyons 1965: ch. 4;
Hare 1963: 130–6). But even if the distinction is valid, it seems
unduly optimistic to assume that utility-maximizing rules will
always protect the rights of weak and unpopular minorities. As
Williams puts it, the assurance that justice will prevail is 'a tribute
to the decency and imagination of those utilitarians but not to their
consistency or their utilitarianism' (Williams 1972: 103). In any
event, this response does not answer the objection, for even if it gets
the right answer, it does so for the wrong reasons. On the rule-
utilitarian view, the wrong done in discriminating against a
minority group is the increased fear caused to others by having a
rule allowing discrimination. The wrong done in not paying the boy
who mowed my lawn is the increased doubts caused in others
concerning the institution of promising. But that is absurd. The
wrong is done to the person who should not have suffered from the
dislike of others, and to the boy who had a special claim to the
promised money. This wrong is present whatever the long-term
effects on others.

The rule-utilitarian response misses the real issue. The objection
to utilitarian decision-making was that certain special obligations
should be included, and that certain illegitimate preferences should
be excluded. These are moral requirements which take precedence
over the maximization of utility (whereas the U-agent sees them
merely as devices for maximizing utility). But if that was our
objection, then it is irrelevant to say, as rule-utilitarians do, that
obeying promises and discounting prejudices often maximizes long-
term utility, or that promises and human rights are even more
ingenious devices for maximizing utility than we initially thought.
That response confirms, rather than refutes, the criticism that U-
agents treat the recognition of special obligations as subject to,
rather than prior to, the maximization of utility. Our objection
was not that promises are bad devices for maximizing utility, but
that they are not such devices at all. This problem cannot be
avoided by changing the level at which we apply the principle of
utility from acts to rules. The problem, from the point of view of
our everyday morality, is in applying the principle of utility itself.

Some utilitarians would agree with what I have said so far. It is
right and proper, they say, to view our attachments as taking
precedence over the pursuit of overall utility. We should accept the
everyday view that the harm done to the particular individuals who

are cheated or discriminated against is sufficient grounds for demanding that people keep promises and respect rights. We should not be U-agents who decide how to act by making utilitarian calculations, and who view promises as devices for maximizing utility. Instead we should view promises, and other people's rights, as of such towering importance that they are basically invulnerable to the calculus of social interests. In short, we should be non-utilitarians in our moral reasoning. But, they argue, this does not mean that utilitarianism is wrong. On the contrary, the reason why we should be non-utilitarians in our decision-making is precisely that we are more likely to maximize utility that way. A society of non-utilitarians who believe in the intrinsic importance of promises and rights will do better, in terms of maximizing utility, than a society of act- or rule-utilitarians who view promises and rights as devices for maximizing utility.

This may sound paradoxical. But it raises a true and important point. Utilitarianism is essentially a 'standard of rightness', not a 'decision-procedure' (Brink 1986: 421–7; Railton 1984: 140–6).[7] What defines utilitarianism is the claim that the right act is the one that maximizes utility, not the claim that we should deliberately seek to maximize utility. It is an open question whether we should employ a utilitarian decision-procedure—indeed, this question is itself to be answered by examining the consequences on overall utility of different decision-procedures. And it is quite possible that we would do better in terms of the utilitarian standard of rightness by employing a non-utilitarian decision-procedure. This certainly seems true in regard to our personal attachments—everyone's life is less valuable if we are unable to make commitments in the sort of whole-hearted and unconditional way precluded by direct utilitarianism. Hence we should be 'indirect utilitarians'.

While this is an important point, it does not yet answer the objections raised above. Consider our everyday view that certain kinds of preferences are unfair, and so should not be given any weight in our moral decision-procedures. It is possible that the utilitarian standard of rightness can justify our adopting such a non-utilitarian decision-procedure. If so, then both sides agree that certain preferences should not be counted. But on our everyday view, the reason why unfair preferences should not be given any weight in our decision-procedure is that they are morally illegiti-mate—they do not deserve to be counted. For the indirect utilitarian, on the other hand, the reason we do not count unfair preferences is simply that it is counter-productive to do so. Unfair preferences (if rational and informed) are as legitimate as any other preference according to the utilarian standard of rightness, but we do better in terms of that standard by treating them as illegitimate in our decision-making.

So we have two conflicting explanations for treating certain preferences as illegitimate. To defend utilitarianism, therefore, it is not enough to show that the utilitarian standard of rightness can justify using non-utilitarian decision-procedures. One must also show that this is the right justification. The utilitarian says that the reason why we use non-utilitarian procedures is that they happen to maximize utility. But is it not more plausible to say that the reason why we use non-utilitarian procedures is simply that we accept a non-utilitarian standard of rightness? Why think there has to be some indirect utilitarian explanation for our non-utilitarian commitments?

Some utilitarians seem to think that if a utilitarian explanation is available for our moral convictions then there is no need to consider any non-utilitarian explanations. But this begs the question. We need some argument for endorsing the utilitarian standard of rightness over alternative standards. Is there any such argument in utilitarian writings? There are in fact two distinct arguments, but I will argue that neither works on its own, and that the plausibility of utilitarianism depends on conflating the two. Once we have examined these arguments, we will see that the problems discussed above stem directly from the utilitarian standard of rightness, and are not substantially affected by how that standard is applied.

4. TWO ARGUMENTS FOR UTILITY-MAXIMIZATION

In this section, I will consider the two main arguments for viewing utility-maximization as the standard of moral rightness (whether or not this standard is employed as a decision-procedure). As we will see, they generate two entirely different intepretations of what utilitarianism is.

(a) Equal consideration of interests

On one interpretation, utilitarianism is a standard for aggregating individual interests and desires. Individuals have distinct and potentially conflicting preferences, and we need a standard that specifies which trade-offs amongst those preferences are morally acceptable, which trade-offs are fair to the people whose welfare is at stake. That is the question which this first interpretation of utilitarianism attempts to answer. One popular answer, found in many different theories, is that each person's interests should be given equal consideration. Each person's life matters equally, from the moral point of view, and hence their interests should be given equal consideration.

Utilitarianism, on this first view of it, accepts this general egalitarian principle. However, the idea of treating people with equal consideration is imprecise, and it needs to be spelled out in more detail if it is to provide a determinate standard of rightness. One obvious, and perhaps initially appealing, way to spell out that idea is to give equal weight to each person's preferences, regardless of the content of the preferences or the material situation of the person. As Bentham put it, we count everyone for one, no one for more than one. On the first account of utilitarianism, then, the reason that we should give equal weight to each person's preferences is that that treats people as equals, with equal concern and respect.

If we accept this as our standard of rightness, then we will conclude that morally right actions are those that maximize utility. But it is important to note that maximization is not the direct goal of the standard. Maximization arises as a by-product of a standard that is intended to aggregate people's preferences fairly. The requirement that we maximize utility is entirely derived from the prior requirement to treat people with equal consideration. So the first argument for utilitarianism is this:

1. people matter, and matter equally; therefore
2. each person's interests should be given equal weight; therefore
3. morally right acts will maximize utility.

This equal consideration argument is implicit in Mill's claim that 'In the golden rule of Jesus of Nazareth, we read the complete spirit

of the ethics of utility. To do as you would be done by, and to love
your neighbour as yourself, constitute the ideal perfection of
utilitarian morality' (Mill 1968: 16). The argument is more
explicitly affirmed by contemporary utilitarians like Harsanyi,
Griffin, Singer, and Hare (Harsanyi 1976: 13–14, 19–20, 45–6, 65–
7; Griffin 1986: 208–15, 295–301; Hare 1984: 106–12; Singer
1979: 12–23; Haslett 1987: 40–3, 220–2). Hare, in fact, finds it
difficult to imagine any other way of showing equal consideration
for each person (Hare 1984: 107; cf. Harsanyi 1976: 35).

(b) Teleological utilitarianism

There is, however, another interpretation of utilitarianism. Here
maximizing the good is primary, not derivative, and we count
individuals equally only because that is the way to maximize value.
Our primary duty is not to treat people as equals, but to bring
about valuable states of affairs. People, as Williams puts it, are just
viewed as *locations* of utilities, or as causal levers for the 'utility
network'. The 'basic bearer of value for Utilitarianism is the *state of
affairs*' (Williams 1981: 4). Utilitarianism, on this view, is primarily
concerned not with persons, but with states of affairs. Rawls calls
this a 'teleological' theory, which means that the right act is defined
in terms of maximizing the good, rather than in terms of equal
consideration for individuals (Rawls 1971: 24).

This second interpretation is a genuinely distinct form of
utilitarianism, not simply a different way of describing the same
theory. Its distinctiveness becomes clear if we look at utilitarian
discussions of population policy. Derek Parfit asks whether we
morally ought to double the population, even if it means reducing
each person's welfare by almost half (since that will still increase
overall utility). He thinks that a policy of doubling the population is
a genuine, if somewhat repugnant, conclusion of utilitarianism. But
it need not be if we view utilitarianism as a theory of treating people
as equals. Non-existent people do not have claims—we do not have
a moral duty to them to bring them into the world. As John Broome
notes, 'one cannot owe anyone a duty to bring her into existence,
because failing in such a duty would not be failing anyone' (Broome
1990–1: 92). So what is the duty here, on the second interpretation?
The duty is to maximize value, to bring about valuable states of

affairs, even if the effect is to make all existing persons worse off than they would otherwise have been.

The distinctness of this second interpretation is also apparent in Thomas Nagel's discussion. He demands that we add a 'deontological' constraint of equal treatment on to utilitarianism, which he thinks is concerned with selecting the 'impersonally best outcome' (Nagel 1986: 176). Nagel says we must qualify our obligation to maximize the good with the obligation to treat people as equals. Obviously his demand only makes sense with reference to the second interpretation of utilitarianism, according to which the fundamental duty is not to aggregate individual preferences fairly, but to bring about the most value in the world. For on the first interpretation, utilitarianism is already a principle of moral equality; if it fails as a principle of equal consideration, then the whole theory fails, for there is no independent commitment to the idea of maximizing utility.

This second interpretation stands the first interpretation on its head. The first defines the right in terms of treating people as equals, which leads to the utilitarian aggregation standard, which happens to maximize the good. The second defines the right in terms of maximizing the good, which leads to the utilitarian aggregation standard, which as a mere consequence treats people's interests equally. As we have seen, this inversion has important theoretical and practical consequences.

So we have two independent, and indeed conflicting, paths to the claim that utility ought to be maximized. Which is the fundamental argument for utilitarianism? Up to this point, I have implicitly relied on the first view—that is, utilitarianism is best viewed as a theory of how to respect the moral claim of each individual to be treated as an equal. Rawls, however, says that utilitarianism is fundamentally a theory of the second sort—i.e. one which defines the right in terms of maximizing the good (Rawls 1971: 27). But there is something bizarre about that second interpretation. For it is entirely unclear why maximizing utility, as our direct goal, should be considered a *moral* duty. To whom is it a duty? Morality, in our everyday view, is a matter of interpersonal obligations—the obligations we owe to each other. But to whom do we owe the duty of maximizing utility? It cannot be to the maximally valuable state of affairs itself, for states of affairs do not have moral claims. Perhaps we have a duty to those people who would benefit from the maximization of utility. But if that duty is, as seems most plausible, the duty to treat people with equal consideration, then we are back to the first interpretation of utilitarianism as a way of treating people as equals. Maximizing utility is now just a by-product, not the ultimate ground of the theory. And then we need not double the population, since we have no obligation to conceive those who would constitute the increased population.

If we none the less accept that maximizing utility is itself the goal, then it is best seen as a non-moral ideal, akin in some ways to an aesthetic ideal. The appropriateness of this characterization can be seen by looking at the other example Rawls gives of a teleologist, namely Nietzsche (Rawls 1971: 25). The good which Nietzsche's theory seeks to maximize (e.g. creativity) is available only to the special few. Others are useful only in so far as they promote the good of the special few. In utilitarianism, the value being maximized is more mundane, something that every individual is capable of partaking of or contributing to (although the maximizing policy may result in the sacrifice of many). This means that in utilitarian teleology, unlike Nietzsche's, every person's preferences must be given some weight. But in neither case is the fundamental principle to treat people as equals. Rather it is to maximize the good. And in both cases, it is difficult to see how this can be viewed as a moral principle. The goal is not to respect *people*, for whom certain things are needed or wanted, but rather to respect the *good*, to which certain people may or may not be useful contributors. If people have become the means for the maximization of the good, morality has dropped out of the picture, and a non-moral ideal is at work. A Nietzschean society may be aesthetically better, more beautiful, but it is not morally better (Nietzsche himself would not have rejected this description—his theory was 'beyond good and evil'). If utilitarianism is interpreted in this teleological way, then it too has ceased to be a moral theory.

I said earlier that one of utilitarianism's attractions was its secular nature—for utilitarians, morality matters because human beings matter. But that attractive idea is absent from this second interpretation, whose moral point is quite obscure. Humans are viewed as potential producers or consumers of a good, and our duties are to that good, not to other people. That violates our core intuition that morality matters because humans matter. In fact, few people have endorsed utilitarianism as a purely teleological theory, without appealing at all to the ideal of equal respect for persons (G. E. Moore's *Ethics* is one prominent exception). Utilitarianism simply ceases to have any attraction if it is cut off from that core intuition.

If utilitarianism is best seen as an egalitarian doctrine, then there is no independent commitment to the idea of maximizing welfare. The utilitarian has to admit that we should use the maximizing standard only if that is the best account of treating people as equals. This is important, because much of the attraction of utilitarianism depends on a tacit mixing of the two justifications.[8] Utilitarianism's intuitive unfairness would quickly disqualify it as an adequate account of equal consideration, were it not that many people take its maximizing feature as an additional, independent reason to endorse it. Utilitarians tacitly appeal to the good-maximization standard to deflect intuitive objections to their account of equal consideration. Indeed, it may seem to be a unique strength of utilitarianism that it can mix these two justifications. Unfortunately, it is incoherent to employ both standards in the same theory. One cannot say that morality is fundamentally about maximizing the good, while also saying that it is fundamentally about respecting the claim of individuals to equal consideration. If utilitarians were held to one or other of the standards, then their theory would lose much of its attractiveness. Viewed as a maximizing-teleological theory, it ceases to meet our core intuitions about the point of morality; viewed as an egalitarian theory, it leads to a number of results which conflict with our sense of what it is to treat people as equals, as I now hope to show in a more systematic way.

5. INADEQUATE CONCEPTION OF EQUALITY

If we are to treat utilitarianism as a plausible political morality, then we must interpret it as a theory of equal consideration. That may seem strange, given the inegalitarian acts utilitarianism might justify—e.g. depriving disliked people of their liberty. But we need to distinguish different levels at which equality can be a value. While utilitarianism may have unequal effects on people, it can none the less claim to be motivated by a concern for treating people as equals. Indeed, Hare asks, if we believe that people's essential interest is the satisfaction of their informed preferences, and that everyone is to be given equal consideration, then what else can we do except give equal weight to each person's preferences, everyone counting for one, no one for more than one (Hare 1984: 106)?

But while utilitarianism seeks to treat people as equals, it violates many of our intuitions about what it genuinely means to treat people with equal consideration. It is possible that our anti-utilitarian intuitions are unreliable. I will argue, however, that utilitarianism has misinterpreted the ideal of equal consideration for each person's interests, and, as a result, it allows some people to be treated as less than equals, as means to other people's ends.

Why is utilitarianism inadequate as an account of equal consideration? Utilitarians assume that every source of happiness, or every kind of preference, should be given the same weight, if it yields equal utility. I will argue that an adequate account of equal consideration must distinguish different kinds of preferences, only some of which have legitimate moral weight.

(a) External preferences

One important distinction amongst kinds of preferences is that between 'personal' and 'external' preferences (Dworkin 1977: 234). Personal preferences are preferences about the goods, resources, and opportunities, etc. one wants available to oneself. External preferences concern the goods, resources, and opportunities one wants available to others. External preferences are sometimes prejudiced. Someone may want blacks to have fewer resources because he thinks them less worthy of respect. Should this sort of external preference be counted in the utilitarian calculus? Does the existence of such preferences count as a moral reason for denying blacks those resources?

As we have seen, indirect utilitarians argue that there are circumstances where we would be better off, in utilitarian terms, by excluding such preferences from our everyday decision-procedures. But the question I want to consider here is whether these preferences should be excluded more systematically, by excluding them from our standard of rightness. And I want to consider whether utilitarianism's own deepest principle provides grounds for not according external preferences any moral weight in its standard of rightness. The deepest principle, we have seen, is an egalitarian one. Each person has an equal moral standing, each person matters as much as any other—that is why each person's preferences should count in the calculus. But if that is why we are attracted to utilitarianism, then it seems inconsistent to count external preferences. For if external preferences are counted, then what I am rightfully owed depends on how others think of me. If they think I am unworthy of equal concern, then I will do less well in the utilitarian aggregation. But utilitarians cannot accept that result, because utilitarianism is premissed on the view that everyone ought to be treated as equals.

If we believe that everyone is to be treated as equals, then it offends our deepest principles to allow some people to suffer because others do not want them treated as equals. As Dworkin puts it, inegalitarian external preferences 'are on the same level— purport to occupy the same space—as the utilitarian theory'. Hence utilitarianism 'cannot accept at once a duty to defeat the false theory that some people's preferences should count for more than other people's and a duty to strive to fulfill the [external] preferences of those who passionately accept that false theory, as energetically as it strives for any other preferences' (Dworkin 1985: 363). The very principle that tells us to count equally every person's preferences in our standard of rightness also tells us to exclude those preferences which deny that people's preferences are to count equally. To paraphrase Harsanyi, utilitarians should be 'conscientious objectors' when faced with such preferences (Harsanyi 1977: 62; Goodin 1982: 93–4).

(b) Selfish preferences

A second kind of illegitimate preference involves the desire for more than one's own fair share of resources. I will call these 'selfish preferences', since they ignore the fact that other people need the resources, and have legitimate claims to them. As with inegalitarian external preferences, selfish preferences are often irrational and uninformed. But satisfying selfish preferences will sometimes generate genuine utility. Should such preferences, if rational, be included in the utilitarian standard of rightness?

Utilitarians will object to the way I have phrased the question. As we have seen, utilitarians deny that there is such a thing as a fair share (and hence a selfish preference) independently of utilitarian calculations. For utilitarians, a fair distribution just is one that maximizes utility, and so no preference can be identified as selfish prior to utility calculations. So it begs the question against utilitarianism to assume that we can identify such things as selfish preferences prior to utilitarian calculations. But we can ask whether the utilitarian's own deepest principle provides grounds for adopting a theory of fair shares that enables us to identify and exclude selfish preferences from our standard of rightness.

This issue is discussed in a recent debate between Hare and John Mackie. Hare, like most utilitarians, believes that all rational preferences should be included in utility aggregation, even those that seem unfair. Even if I have a massive amount of resources, while my neighbour has very little, if I covet my neighbour's resources, then my desire must be included in the calculation. And if the calculations work out in my favour, perhaps because I have many friends who would share in my enjoyment, then I should get those resources. No matter how much I already have, my desire for more resources continues to count equally, even when the resources I want must come from someone with very little.

Why should utilitarians count such preferences? Hare believes that the principle of equal consideration requires it. According to Hare, the best way to interpret that egalitarian principle is to use the following mental test: we put ourselves in other people's shoes, and try to imagine how our actions affect them. And we should do this for everyone affected by our actions. We take the viewpoint of each person and treat it as being equally important as our own viewpoint, equally worthy of concern. Indeed, Hare says, we should treat these other viewpoints *as* our own viewpoint. This ensures that we are showing equal consideration for each person. If we have, in this way, put ourselves in everyone else's shoes, then we should choose that action which is best for 'me', where 'me' here means all of the 'mes', i.e. all of the different viewpoints I am now considering as equally my own. If I try to choose what is best for all my different selves, I will choose that action which maximizes the preference satisfaction of all these 'selves'. So, Hare claims, the utilitarian aggregation criterion follows naturally from this in-tuitive model of equal consideration. If I treat each person's interests as mattering equally, by imagining that their viewpoint is in fact one of my own, then I will adopt utilitarian principles (Hare 1984: 109–10; cf. 1982: 25–7).

Hare thinks that this is the only rational way of showing equal concern for people. But as Mackie notes, there are other possibilities, even if we accept Hare's claim that we treat people as equals by putting ourselves in their shoes, and treating each of these different selves as equally important. Rather than maximize preference-satisfaction amongst all these selves, we might show our concern for them by guaranteeing each 'a fair go' in life, i.e. guaranteeing each an adequate level of resources and liberties. Or we might, when successively occupying these different positions, do what is best for the least well off, or provide each with an equal share of the available resources and liberties. These are all different conceptions of what the abstract notion of equal consideration requires (Mackie 1984: 92).

How can we decide between these different ways of showing equal consideration? Utilitarians point out that their view may also lead to an egalitarian distribution of resources. People who lack resources will, in general, get more utility out of each additional resource than those who already have many resources. Someone who is starving is sure to get more utility from a piece of food than someone who is already well supplied with food (Hare 1978: 124–6; Brandt 1959: 415–20). So both sides can agree to start with a roughly equal distribution of resources. However, Hare and Mackie conceive this initially equal distribution in very different ways.

For Mackie, so long as everyone else has their fair share of resources, then the resources initially allotted to me are mine—i.e. no one else has any legitimate claim of justice over them. Some people who already have their fair share may also want some of my share. But that is not important, morally speaking. Their preferences have no moral weight. They are selfish preferences, since they fail to respect my claim to a fair share. On Mackie's view, the state should secure each person's share of resources, and not allow them to be taken away just because other people have selfish preferences for what is rightfully someone else's. The best conception of equal consideration would exclude such selfish preferences.

For Hare, on the other hand, the resources initially distributed to me are not really mine in the same way. They are mine unless or until someone else can make better use of them, where 'better' means more productive of overall utility. Hare thinks this proviso for taking away my share is required by the same value that led the government initially to give it to me, i.e. an equal concern for each person's goals. If we care equally about people's goals, then it is right to redistribute resources whenever we can satisfy more goals by so doing.

Do we have any reason to choose one of these conceptions of equal consideration over the other? We need to look more closely at the kinds of preferences that would be involved in Hare's redistribution. Let us assume that I have my fair share, as does everyone else, and that we are in an affluent society, so that this share includes a house and lawn. Everyone else on my block plants a flower garden, but they would like my lawn left open as a public space for children to play on, or to walk dogs on. I, however, want my own garden. The desires of others to use my lawn as a public space may well outweigh, in terms of overall utility, my desire to have a garden. Hare thinks it is right, therefore, to sacrifice my desire for the greater desires of others.

If it is morally wrong for me to insist on having a garden, we need to know *who is wronged*. If my sacrifice is required to treat people as equals, who is treated as less than an equal if I disallow the sacrifice? Hare's answer is that the other members of the block are not treated as equals if their preferences are not allowed to outweigh my desire. But surely that is implausible, since they already have their own yard, their own fair share of resources. According to Hare, my neighbours' desire to decide how to use my resources, as well as their own, is a legitimate preference which grounds a moral claim. But is it not more accurate to describe such a preference as simply selfish? Why should my neighbours suppose that the idea of equal concern gives them any claim over my share of resources? If they already have their own lawn, then I am not treating them unjustly in saying that my preference concerning my lawn outweighs or pre-empts their preferences. I still respect them as equals since I make no claim on the resources they have to lead their lives. But they do not respect me as an equal when they expect or demand that I give up my share of resources to satisfy their selfish desire to have more than their fair share.

This points to an important component of our everyday sense of what it means to treat people as equals—namely, we should not expect others to subsidize our projects at the expense of their own. Perhaps my friends and I have expensive tastes—we like to eat caviar and play tennis all day. To expect others to give up their fair share of resources to support our taste, no matter how happy it makes us, is selfish. If I already have my share of resources, then to suppose that I have a legitimate moral claim to someone else's resources, just because it will make me happier, is a failure to show equal concern for others. If we believe that others should be treated as equals, then we will exclude such selfish preferences from the utilitarian calculus.

So the very principle which supported an initially equal distribution of resources also argues for securing that distribution. Hare's proviso—that the initial distribution be subject to utility-maximizing redistribution—undermines, rather than extends, the point of the initial distribution. Hare's idea of treating other people's interests as my own when engaged in moral reasoning is not necessarily a bad one. It is one way of rendering vivid the idea of moral equality (we will look at other such devices in the next chapter). But the equal concern he seeks to promote is not achieved by treating other people's preferences as constituting equal claims on all of our actions and resources. Rather, equality teaches us how much by way of resources we have to pursue our projects, and how much is rightfully left for others. Equal concern is shown by ensuring that others can claim their own fair share, not by ensuring that they have equal weight in determining the use of my share. Securing people's fair shares, rather than leaving them subject to selfish preferences, is the better spelling out of the equal concern that Hare seeks.

This, according to Rawls, is a fundamental difference between his account of justice and the utilitarians'. For Rawls, it is a defining feature of our sense of justice that 'interests requiring the violation of justice have no value', and so the presence of illegitimate preferences 'cannot distort our claims upon one another' (Rawls 1971: 31, 450, 564). Justice 'limits the admissible conceptions of the good, so that those conceptions the pursuit of which violate the principles of justice are ruled out absolutely: the claims to pursue inadmissible conceptions have no weight at all'. Because unfair preferences 'never, so to speak, enter into the social calculus', people's claims 'are made secure from the unreasonable demands of others'. For utilitarians, on the other hand, 'no restrictions founded on right and justice are imposed on the ends through which satisfaction is to be achieved' (Rawls 1982*b*: 184, 171 n., 170, 182).

We can now see why utilitarianism fails to recognize special relationships, or to exclude illegitimate preferences. In each case, utilitarianism is interpreting equal consideration in terms of the aggregation of pre-existing preferences, whatever they are for, even if they invade the rights or commitments of others. But our intuitions tell us that equality should enter into the very formation of our preferences. Part of what it means to show equal consideration for others is taking into account what rightfully belongs to them in deciding on one's own goals in life.[9] Hence prejudiced and selfish preferences are excluded from the start, for they already reflect a failure to show equal consideration. However, if my goals do respect other people's rightful claims, then I am free to pursue special relationships, even if some other act maximizes utility. If my plans respect the teachings of equality, then there is nothing wrong with giving priority to my family or career. This means that my day-to-day activities will show unequal concern—I will care more about helping my friends, or the causes I am committed to, than about helping the goals of other people. That is part of what it means to have friends and causes. And that is entirely acceptable, so long as I respect the claims of others concerning the pursuit of their projects.

If we think about the values that motivate utilitarianism, the values which give it its initial plausibility, we will see that it must be modified. Utilitarianism is initially attractive because human beings matter and matter equally. But the goal of equal consideration that utilitarians seek to implement is best implemented by an approach that includes a theory of fair shares. Such a theory would exclude prejudiced or selfish preferences that ignore the rightful claims of others, but would allow for the kinds of special commitments that are part of our very idea of leading a life. These modifications do not conflict with the general principle of consequentialism, but rather stem from it. They are refinements of the general idea that morality should be about the welfare of human beings. Utilitarianism has simply over-simplified the way in which we intuitively believe that the welfare of others is worthy of moral concern.

As we have seen, indirect utilitarians claim that our intuitive commitment to non-utilitarian decision-procedures does not undermine utilitarianism as a standard of rightness, since we can give a utilitarian justification for adopting non-utilitarian procedures. But that response will not work here, for my argument concerns utilitarianism as a standard of rightness. My claim is that the very reason utilitarians give for basing their standard of rightness on the satisfaction of people's preferences is also a reason to exclude external and selfish preferences from that standard. This is an objection to the theory's principles, not to the way those principles get applied in decision-procedures.

Commentators who endorse these sorts of modifications of utilitarianism often describe the resulting theory as a balance or compromise between the values of utility and equality (e.g. Raphael 1981: 47–56; Brandt 1959: ch. 16; Hospers 1961: 426; Rescher 1966: 59). That is not what I have argued. Rather, the modifications are needed to provide a better spelling out of the ideal of equal consideration which utilitarianism itself appeals to.

It is worth pausing to consider the kind of argument that I have just presented, since it expresses, I believe, one basic form of political argument. As I mentioned in the introduction, the idea of equality is often said to be the basis of political morality. Both Hare's utilitarianism and Mackie's 'right to a fair go' appeal to the idea that each person is entitled to equal consideration. But they do not give an equally compelling account of that idea. Our intuitions tell us that utilitarianism fails to ensure that people are treated as equals, since it lacks a theory of fair shares.

This might suggest that political theorizing is a matter of correctly deducing specific principles from this shared premiss of moral equality. Political argument, then, would primarily be a matter of identifying mistaken deductions. But political philosophy is not like logic, where the conclusion is meant to be already fully present in the premisses. The idea of moral equality is too abstract for us to be able to deduce anything very specific from it. There are many different and conflicting kinds of equal treatment. Equality of opportunity, for example, may produce unequal income (since some people have greater talents), and equal income may produce unequal welfare (since some people have greater needs). All of these particular forms of equal treatment are logically compatible with the idea of moral equality. The question is which form of equal treatment best captures that deeper ideal of treating people as equals. That is not a question of logic. It is a moral question, whose answer depends on complex issues about the nature of human beings and their interests. In deciding which particular form of equal treatment best captures the idea of treating people as equals, we do not want a logician, who is versed in the art of logical deductions. We want someone who has an understanding of what it is about humans that deserves respect and concern, and of what kinds of activities best manifest that respect and concern.

The idea of moral equality, while fundamental, is too abstract to serve as a premiss from which we deduce a theory of justice. What we have in political argument is not a single premiss and then competing deductions, but rather a single concept and then competing conceptions or interpretations of it. Each theory of justice is not *deduced from* the ideal of equality, but rather *aspires to* it, and each theory can be judged by how well it succeeds in that aspiration. As Dworkin puts it, when we instruct public officials to act in accordance with the concept of equality, we 'charge those whom [we] instruct with the responsibility of developing and applying their own conception. . . . That is not the same thing, of course, as granting them a discretion to act as they like; it sets a standard which they must try—and may fail—to meet, because it assumes that one conception is superior to another' (Dworkin 1977: 135).[10] However confident we are in a particular conception of equality, it must be tested against competing conceptions to see which best expresses or captures the concept of equality.

This is the kind of argument I have tried to give against utilitarianism. We can see the weakness in utilitarianism as a conception of equality by comparing it to a conception which guarantees certain rights and fair shares of resources. When we compare these two conceptions, utilitarianism seems implausible as an account of moral equality, at odds with our intuitions about that basic concept. But its implausibility is not a matter of logical error, and the strength of a theory of fair shares is not a matter of logical proof. This may be unsatisfying to those accustomed to more rigorous forms of argument. But if the egalitarian suggestion is correct—if each of these theories is aspiring to live up to the ideal of treating people as equals—then this is the form that political argument must take. To demand that it achieve logical proof simply misunderstands the nature of the exercise. Any attempt to spell out and defend our beliefs about the principles which should govern the political community will take this form of comparing different conceptions of the concept of equality.

6. THE POLITICS OF UTILITARIANISM

What are the practical implications of utilitarianism as a political morality? I have argued that utilitarianism could justify sacrificing the weak and unpopular members of the community for the benefit of the majority. But utilitarianism has also been used to attack those who hold unjust privileges at the expense of the majority. Indeed, utilitarianism, as a self-conscious political and philosophical movement, arose as a radical critique of English society. The original utilitarians were 'Philosophical Radicals' who believed in a complete rethinking of English society, a society whose practices they believed were the product not of reason, but of feudal superstition. Utilitarianism, at that time, was identified with a progressive and reform-minded political programme—the extension of democracy, penal reform, welfare provisions, etc.

Contemporary utilitarians, on the other hand, are 'surprisingly conformist'—in fact they seem keen to show that utilitarianism leaves everything as it is (Williams 1972: 102). Whereas the original utilitarians were willing to judge existing social codes at the altar of human well-being, many contemporary utilitarians argue there are good utilitarian reasons to follow everyday morality uncritically. It may seem that we can increase utility by making exceptions to a rule of everyday morality, but there are utilitarian reasons for sticking to good rules under all circumstances. And even if it seems that the everyday rule is not a good one in utilitarian terms, there are utilitarian reasons for not evaluating rules in terms of utility. It is difficult to predict the consequences of our actions, or to measure these consequences even when known. Hence our judgements about what maximizes utility are imperfect, and attempts to rationalize social institutions are likely to cause more harm than good. The gains of new rules are uncertain, whereas existing conventions have proven value (having survived the test of cultural evolution), and people have formed expectations around them. Moreover, acting directly on utilitarian grounds is counterproductive, for it encourages a contingent and detached attitude towards what should be whole-hearted personal and political commitments.

As a result, modern utilitarians downplay the extent to which utilitarianism should be used as a critical principle, or as a principle of political evaluation at all. Some utilitarians say we should only resort to utilitarian reasoning when our everyday precepts lead to conflicting results; others say that the best world, from a utilitarian point of view, is one in which no one ever reasons in an explicitly utilitarian manner. Williams claims that this sort of utilitarianism is self-defeating—it argues for its own disappearance. This is not self-defeating in the technical sense, for it does not show that the morally right action is not, after all, the one that maximizes utility. But it does show that utilitarianism is no longer being offered as the correct language for political debate. Politics should be debated in the non-utilitarian language of everyday morality—the language of rights, personal responsibilities, the public interest, distributive justice, etc. Utilitarianism, on some modern views of it, leaves everything as it is—it stands above, rather than competes with, everyday political decision-making.

Some utilitarians continue to claim that utilitarianism requires a radical critique of the arbitrary and irrational aspects of everyday morality (e.g. Singer 1979). But it is unlikely that utilitarianism will ever form a coherent political movement, such as characterized its birth. The problem is that 'the winds of utilitarian argumentation blow in too many directions' (Sher 1975: 159). For example, while some utilitarians argue that utility is maximized by massive redistribution of wealth, due to the declining marginal utility of money, others defend *laissez-faire* capitalism because it creates more wealth. This is not just a question of predicting how different economic policies fare in terms of an agreed-upon scale of utility. It is also a question about how to define the scale—what is the relationship between economic goods and other components of the human good (leisure, community, etc.)? It is also a question of the role of utility calculations themselves—how reliably can we determine overall utility, and how important are established conventions? Given these disagreements about how and when to measure utility, utilitarianism is bound to yield fundamentally opposed judgements.

I do not mean to suggest that all these positions are equally plausible (or that these problems are not also found in non-utilitarian theories). The confidence and unanimity that the original utilitarians had in their political judgements was often the result of an over-simplified view of the issues, and a certain amount of indeterminacy is unavoidable in any theory once we recognize the complexity of the empirical and moral issues involved. Modern utilitarians are right to insist that utility is not reducible to pleasure, and that not all kinds of utility are measurable or commensurable, and that it is not always appropriate even to try to measure these utilities. However, the price of this added sophistication is that utilitarianism does not immediately identify any set of policies as distinctly superior. Modern utilitarianism, despite its radical heritage, no longer defines a distinctive political position.

NOTES

1. This common slogan is misleading because it contains two distinct maximands—'greatest happiness' and 'greatest number'. It is impossible for any theory to contain a double maximand, and any attempt to implement it quickly leads to an impasse (e.g. if the two possible distributions are 10 : 10 : 10 and 20 : 20 : 0, then we cannot produce both the greatest happiness and the happiness of the greatest number). See Griffin (1986: 151–4); Rescher (1966: 25–8).
2. Of course, while I might prefer A if informed, it does not follow that A provides me with any benefit in my current uninformed state. This complicates the informed preference account of utility, but does not subvert it. What promotes my well-being is distinct from satisfying my existing preferences, even if it is also distinct from satisfying my ideally informed preferences (Griffin 1986: 11–12; 32–3). It is possible, however, that a full development of this account would bring it close to what is sometimes called an 'Objective List' theory (Parfit 1984: 493–502).
3. I do not believe that the preferences of the dead are always without moral weight. What happens after our death can affect how well our life went, and our desire for certain things after our death can be an important focus for our activities in life. Indeed, if the preferences of the dead did not sometimes have moral weight, it would be impossible to make sense of the way we treat wills. See the discussion in Lomasky (1987: 212–21), and Feinberg (1980: 173–6). On the 'experience requirement' more generally, see Larmore (1987: 48–9), Lomasky (1987: 231–3), Griffin (1986: 13–23), Parfit (1984: 149–53).
4. Political theories which are concerned with the distribution of resources, without determining the effect these resources have on each person's welfare, may seem an exception to this general claim. But, as I will discuss in chapter 3, this is a misleading perception, and even

resource-based theories must have some theory of people's essential interests 'most comprehensively construed' (Dworkin 1983: 24).

5. It is not clear whether utilitarianism can in fact limit itself to the basic structure of society, or to political decision-making. Even if utilitarianism applies in the first instance to political decisions or social institutions, and not to the personal conduct of individuals, one of the decisions governments face is to determine the legitimate scope of private attachments. If people are not maximizing utility in their private lives, then reorganizing the basic structure so as to leave less room for private life could increase utility. If comprehensive moral utilitarianism cannot accommodate our sense of the value of personal attachments, then political utilitarianism will have no reason to preserve a private realm. In any event, the predominance of utilitarianism in political philosophy stems mostly from the belief that it is the only coherent or systematic moral philosophy (Rawls 1971: vii–viii), and so the motivation for political utilitarianism is undermined if comprehensive moral utilitarianism can be shown to be indefensible.

6. The U-agent is often described as an 'act-utilitarian', because he acts directly on the basis of utility calculations. But this is misleading in so far as 'act-utilitarian' is commonly contrasted with 'rule-utilitarian'. What defines the U-agent is that he uses utility-maximization *directly* as a decision-procedure, and, as we will see, he could do this while focusing on rules rather than acts. The distinction between direct and indirect utilitarianism cuts across the distinction between act- and rule-utilitarianism (Railton 1984: 156–7). The first contrast is whether the principle of utility-maximization is viewed as a decision-procedure or a standard of rightness, not whether the principle of utility-maximization (as either a standard of rightness or a decision-procedure) applies to acts or rules.

7. While the distinction between standards of rightness and decision-procedures is sound, it is not clear that we can make the sort of distinction between them that is required by indirect utilitarianism. Unlike the rule-utilitarian, who views promises as ingenious devices to maximize utility, the indirect utilitarian views our *beliefs about promises* as ingenious devices for maximizing utility. But people do not, and arguably cannot, view their moral beliefs this way (Smith 1988). Moreover, if we put too much weight on the distinction, it is not clear why utilitarianism as a standard of rightness should not disappear entirely from our conscious beliefs (Williams 1973: 135).

8. Critics of utilitarianism also conflate the two versions. This is true, for example, of Rawls's claim that utilitarians ignore the separateness of persons. According to Rawls, utilitarians endorse the principle of maximizing utility because they generalize from the one-person case (it

is rational for each individual to maximize her happiness) to the many-person case (it is rational for society to maximize its happiness). Rawls objects to this generalization because it treats society as if it were a single person, and so ignores the difference between trade-offs within one person's life and trade-offs across lives (Rawls 1971: 27; cf. Nozick 1974: 32–3; Gordon 1980: 40; Mackie 1984: 86–7). However, neither the egalitarian nor the teleological version of utilitarianism makes this generalization, and Rawls's claim rests on a conflation of the two. On this, see Kymlicka (1988*b*: 182–5).

9. This is only part of what equality requires, for there are obligations to those who are unable to help themselves, and Good Samaritan obligations to those who are in dire need. In these cases, we have obligations that are not tied to respecting people's rightful claims. I return to these issues in chapter 7.

10. This shows why it is wrong to claim that Dworkin's egalitarian plateau is 'purely formal' or 'empty' since it is compatible with many different kinds of distributions (Hart 1979: 95–6; Goodin 1982: 89–90; Mapel 1989: 54; Larmore 1987: 62; Raz 1986: ch. 9). As Dworkin notes, this objection 'misunderstands the role of abstract concepts in political theory and debate' (Dworkin 1977: 368). The idea of treating people as equals is abstract, but not formal—on the contrary, it is a substantive ideal that excludes some theories (e.g. racist ones), and that sets a standard to which other theories aspire. The fact that an abstract concept needs to be interpreted, and that different theories interpret it in different ways, does not show that the concept is empty, or that one interpretation of that concept is as good as any other.

3

Liberal Equality

1. RAWLS'S PROJECT

(a) Intuitionism and utilitarianism

In the last chapter I argued that we need a theory of fair shares prior to the calculation of utility, for there are limits to the way individuals can be legitimately sacrificed for the benefit of others. If we are to treat people as equals, we must protect them in their possession of certain rights and liberties. But which rights and liberties?

Most of the political philosophy written in the last twenty years has been on this question. There are some people, as we have seen, who continue to defend utilitarianism. But there has been a marked shift away from the 'once widely accepted old faith that some form of utilitarianism, if only we could discover the right form, *must* capture the essence of political morality' (Hart 1979: 77), and most contemporary political philosophers have hoped to find a systematic alternative to utilitarianism. John Rawls was one of the first to present such an alternative in his 1971 book *A Theory of Justice*. Many others had written about the counter-intuitive nature of utilitarianism. But Rawls starts his book by complaining that political theory was caught between two extremes: utilitarianism on the one side, and an incoherent jumble of ideas and principles on the other. Rawls calls this second option 'intuitionism', an approach which is little more than a series of anecdotes based on particular intuitions about particular issues.

Intuitionism is an unsatisfying alternative to utilitarianism, for while we do indeed have anti-utilitarian intuitions on particular issues, we also want an alternative theory which makes sense of those intuitions. We want a theory which shows why these particular examples elicit disapproval in us. But 'intuitionism' never gets beyond, or underneath, these initial intuitions to show how they are related, or to provide principles that underlie and give structure to them.

Rawls describes intuitionist theories as having two features:

first, they consist of a plurality of first principles which may conflict to give contrary directives in particular types of cases; and second, they include no explicit method, no priority rules, for weighing these principles against one another: we are simply to strike a balance by intuition, by what seems to us most nearly right. Or if there are priority rules, these are thought to be more or less trivial and of no substantial assistance in reaching a judgement. (1971: 34)

There are many kinds of intuitionism, which can be distinguished by the level of generality of their principles.

Common sense intuitionism takes the form of groups of rather specific precepts, each group applying to a particular problem of justice. There is a group of precepts which applies to the question of fair wages, another to that of taxation, still another to punishment, and so on. In arriving at the notion of a fair wage, say, we are to balance somehow various competing criteria, for example, the claims of skill, training, effort, responsibility, and the hazards of the job, as well as to make some allowance for need. No one presumably would decide by any one of these precepts alone, and some compromise between them must be struck. (1971: 35)

But the various principles can also be of a much more general nature. Thus it is common for people to talk about intuitively balancing equality and liberty, or equality and efficiency, and these principles would apply to the entire range of a theory of justice (1971: 36–7). These intuitionist approaches, whether at the level of specific precepts or general principles, are not only theoretically unsatisfying, but are also quite unhelpful in practical matters. For they give us no guidance when these specific and irreducible precepts conflict. Yet it is precisely when they conflict that we look to political theory for guidance.

It is important, therefore, to try to establish some priority amongst these conflicting precepts. This is the task Rawls sets himself—to develop a comprehensive political theory that structures our different intuitions. He does not assume that there is such a theory, but only that it is worth trying to find one:

Now there is nothing intrinsically irrational about this intuitionist doctrine. Indeed, it may be true. We cannot take for granted that there must be a complete derivation of our judgements of social justice from recognizable ethical principles. The intuitionist believes to the contrary that the complexity of the moral facts defies our efforts to give a full account of our judgements and necessitates a plurality of competing principles. He contends that attempts to go beyond these principles either reduce to triviality, as when it is said that social justice is to give every man his due, or else lead to falsehood and oversimplification, as when one settles everything by the principle of utility. The only way therefore to dispute intuitionism is to set forth the recognizably ethical criteria that account for the weights which, in our considered judgements, we think appropriate to give to the plurality of principles. A refutation of intuitionism consists in presenting the sort of constructive criteria that are said not to exist. (1971: 39)

Rawls, then, has a certain historical importance in breaking the intuitionism–utilitarianism deadlock. But his theory is important for another reason. His theory dominates the field, not in the sense of commanding agreement, for very few people agree with all of it, but in the sense that later theorists have defined themselves in opposition to Rawls. They explain what their theory is by contrasting it with Rawls's theory. We will not be able to make sense of later work on justice if we do not understand Rawls.

(b) The principles of justice

In presenting Rawls's ideas, I will first give his answer to the question of justice, and then discuss the two arguments he gives for that answer. His 'general conception of justice' consists of one central idea: 'All social primary goods— liberty and opportunity, income and wealth, and the bases of self-respect—are to be distributed equally unless an unequal distribution of any or all of these goods is to the advantage of the least favoured' (1971: 303). In this 'general conception', Rawls ties the idea of justice to an equal share of social goods, but he adds an important twist. We treat people as equals by removing not all inequalities, but only those which disadvantage someone. If certain inequalities benefit everyone, by drawing out socially useful talents and energies, then they will be acceptable to everyone. If giving someone else more money than I have promotes my interests, then equal concern for my interests suggests that we allow, rather than prohibit, that inequality. Inequalities are allowed if they improve my initially equal share, but are not allowed if, as in utilitarianism, they invade my fair share. That is the single, simple idea at the heart of Rawls's theory.

However, this general conception is not yet a full theory of justice, for the various goods being distributed according to that principle may conflict. For example, we might be able to increase someone's income by depriving them of one of their basic liberties. This unequal distribution of liberty favours the least well off in one way (income) but not in another (liberty). Or what if an unequal distribution of income benefits everyone in terms of income, but creates an inequality in opportunity which disadvantages those with less income? Do these improvements in income outweigh disadvantages in liberty or opportunity? The general conception leaves these questions unresolved, and so does not solve the problem which made intuitionist theories unhelpful.

We need a system of priority amongst the different elements in the theory. Rawls's solution is to break down the general conception into three parts, which are arranged according to a principle of 'lexical priority'.

First Principle—Each person is to have an equal right to the most extensive total system of equal basic liberties compatible with a similar system of liberty for all.

Second Principle—Social and economic inequalities are to be arranged so that they are both:

(*a*) to the greatest benefit of the least advantaged, and
(*b*) attached to offices and positions open to all under conditions of fair equality of opportunity.

First Priority Rule (The Priority of Liberty)—The principles of justice are to be ranked in lexical order and therefore liberty can be restricted only for the sake of liberty.

Second Priority Rule (The Priority of Justice over Efficiency and Welfare)—

The second principle of justice is lexically prior to the principle of efficiency and to that of maximizing the sum of advantages; and fair opportunity is prior to the difference principle. (1971: 302–3)

These principles form the 'special conception' of justice, and they seek to provide the systematic guidance that intuitionism could not give us. According to these principles, some social goods are more important than others, and so cannot be sacrificed for improvements in those other goods. Equal liberties take precedence over equal opportunity which takes precedence over equal resources. But within each category Rawls's simple idea remains—an inequality is only allowed if it benefits the least well off. So the priority rules do not affect the basic principle of fair shares that remains within each category.

These two principles are Rawls's answer to the question of justice. But we have not yet seen his arguments for it. In this chapter, I will focus on Rawls's arguments for the second principle—which he calls the 'difference principle'—governing the distribution of economic resources. I will not discuss the liberty principle, or why Rawls gives priority to it, until later chapters. However, it is important to note that Rawls is not endorsing a general principle of liberty, such that anything that can plausibly be called a liberty is to be given overriding priority. Rather, he is giving special protection to what he calls the 'basic liberties', by which he means the standard civil and political rights recognized in liberal democracies—the right to vote, to run for office, due process, free speech, mobility, etc. (1971: 61). These rights are very important to liberalism—indeed, one way of differentiating liberalism just is that it gives priority to the basic liberties.

However, the assumption that civil and political rights should have priority is widely shared in our society. As a result, the disputes between Rawls and his critics have tended to be on other issues. The idea that people should have their basic liberties protected is the least contentious part of his theory. But my rejection of utilitarianism was based on the need for a theory of fair shares in economic resources as well, and that is more controversial. Some people reject the idea of a theory of fair shares of economic resources, and those who accept it have very different views about what form such a theory should take. This question of resource distribution is central to the shift from utilitarianism to the other theories of justice we will be examining. So I will concentrate for now on Rawls's account of the difference principle.

Rawls has two arguments for his principles of justice. One is to contrast his theory with what he takes to be the prevailing ideology concerning distributive justice—namely, the ideal of equality of opportunity. He argues that his theory better fits our considered intuitions concerning justice, and that it gives a better spelling out of the very ideals of fairness that the prevailing ideology appeals to. The second argument is quite different. Rawls argues that his principles of justice are superior because they are the outcome of a hypothetical social contract. He claims that if people in a certain kind of pre-social state had to decide which principles should govern their society, they would choose his principles. Each person in what Rawls calls the 'original position' has a rational interest in adopting Rawlsian principles for the governing of social co-operation. This second argument has received the most critical attention, and is the one which Rawls is most famous for. But it is not an easy argument to interpret, and we can get a better handle on it if we begin with the first argument.[1]

2. THE INTUITIVE EQUALITY OF OPPORTUNITY ARGUMENT

The prevailing justification for economic distribution in our society is based on the idea of 'equality of opportunity'. Inequalities of income and prestige etc. are assumed to be justified if and only if there was fair competition in the awarding of the offices and positions that yield those benefits. It is acceptable to pay someone a $100,000 salary when the national average is $20,000 if there was fair equality of opportunity—that is, if no one was disadvantaged by their race, or sex, or social background. Such an unequal income is just regardless of whether or not the less well off benefit from that inequality. (This is what Mackie meant by a 'right to a fair go'—see ch. 2, s. 5b above.)

This conflicts with Rawls's theory, for while Rawls also requires equality of opportunity in allotting positions, he denies that the people who fill the positions are thereby entitled to a greater share of society's resources. A Rawlsian society may pay such people more than average, but only if it benefits all members of society to do so. Under the difference principle, people only have a claim to a greater share of resources if they can show that it benefits those who have lesser shares.

Why does the ideology of equal opportunity seem fair to many people in our society? Because it ensures that people's fate is determined by their choices, rather than their circumstances. If I am pursuing some personal ambition in a society that has equality of opportunity, then my success or failure will be determined by my performance, not by my race or class or sex. In a society where no one is privileged or disadvantaged by their social circumstances, people's success (or failure) will be the result of their own choices and efforts. Hence whatever success we achieve is 'earned', rather than merely endowed on us. In a society that has equality of opportunity, unequal income is fair, because success is 'merited', it goes to those who 'deserve' it.

People disagree about what is needed to ensure fair equality of opportunity. Some people believe that legal non-discrimination in education and employment is sufficient. Others argue that affirmative action programmes are required for economically and culturally disadvantaged groups, if their members are to have a genuinely equal opportunity to acquire the qualifications necessary for economic success. But the central motivating idea in each case is this: it is fair for individuals to have unequal shares of social goods if those inequalities are earned and deserved by the individual, that is, if they are the product of the individual's actions and choices. But it is unfair for individuals to be disadvantaged or privileged by arbitrary and undeserved differences in their social circumstances.

Rawls recognizes the attraction of this view. But there is another source of undeserved inequality which it ignores. It is true that social inequalities are undeserved, and hence it is unfair for one's fate to be made worse by that undeserved inequality. But the same thing can be said about inequalities in natural talents. No one deserves to be born handicapped, or with an IQ of 140, any more than they deserve to be born into a certain class or sex or race. If it is unjust for people's fate to be influenced by the latter factors, then it is unclear why the same injustice is not equally present when people's fate is determined by the former factors. The injustice in each case is the same—distributive shares should not be influenced by factors which are arbitrary from the moral point of view. Natural talents and social circumstances are both matters of brute luck, and people's moral claims should not depend on brute luck.

Hence the prevailing ideal of equality of opportunity is 'unstable', for 'once we are troubled by the influence of either social contingencies or natural chance on the determination of distributive shares, we are bound, on reflection, to be bothered by the influence of the other. From a moral standpoint the two seem equally arbitrary' (1971: 74–5). In fact, Dworkin says that the undeserved character of natural assets makes the prevailing view not so much unstable as 'fraudulent' (Dworkin 1985: 207). The prevailing view suggests that removing social inequalities gives each person an equal opportunity to acquire social benefits, and hence suggests that any differences in income between individuals are earned, the product of people's effort or choices. But the naturally handicapped do not have an equal opportunity to acquire social benefits, and their lack of success has nothing to do with their choices or effort. If we are genuinely interested in removing undeserved inequalities, then the prevailing view of equality of opportunity is inadequate.

The attractive idea at the base of the prevailing view is that people's fate should be determined by their choices—by the decisions they make about how to lead their lives—not by the circumstances which they happen to find themselves in. But the prevailing view only recognizes differences in social circumstances, while ignoring differences in natural talents (or treating them as if they were one of our choices). This is an arbitrary limit on the application of its own central intuition.

How should we treat differences in natural talents? Some people, having considered the parallels between social and natural inequality, assume that no one should benefit from their natural inequalities. But as Rawls says, while

no one deserves his greater natural capacity nor merits a more favorable starting place in society . . . it does not follow that one should eliminate these distinctions. There is another way to deal with them. The basic structure can be arranged so that these contingencies work for the good of the least fortunate. Thus we are led to the difference principle if we wish to set up the social system so that no one gains or loses from his arbitrary place in the distribution of natural assets or his initial position in society without giving or receiving compensating advantages in return. (1971: 102)

While no one should suffer from the influence of undeserved natural inequalities, there may be cases where everyone benefits

Liberal Equality

from allowing such an influence. No one deserves to benefit from their natural talents, but it is not unfair to allow such benefits when they work to the advantage of those who were less fortunate in the 'natural lottery'. And this is precisely what the difference principle says.

This is Rawls's first argument for his theory of fair shares. Under the prevailing view, talented people can naturally expect greater income. But since those who are talented do not deserve their advantages, their higher expectations 'are just if and only if they work as part of a scheme which improves the expectations of the least advantaged members of society' (1971: 75). So we get to the difference principle from an examination of the prevailing view of equality of opportunity. As Rawls puts it, 'once we try to find a rendering of [the idea of equality of opportunity] which treats everyone equally as a moral person, and which does not weight men's share in the benefits and burdens of social co-operation according to their social fortune or their luck in the natural lottery, it is clear that the [difference principle] is the best choice among the . . . alternatives' (1971: 75).

That is the first argument. I think the basic premiss of the argument is correct. The prevailing view of equality of opportunity is unstable, and we should recognize that our place in the distribution of natural talents is morally arbitrary. But the conclusion is not quite right. From the fact that natural and social inequalities are arbitrary, it might follow that those kinds of inequalities should only influence distribution when it would benefit the least well off. But the difference principle says that all inequalities must work to the benefit of the least well off. What if I was not born into a privileged social group, and was not born with any special talents, and yet by my own choices and effort have managed to secure a larger income than others? Nothing in this argument explains why the difference principle applies to all inequalities, rather than just to those inequalities which stem from morally arbitrary factors. I will return to this point after examining the second argument.

3. THE SOCIAL CONTRACT ARGUMENT

Rawls considers the first argument for his principles of justice less important than the second. His main argument is a 'social contract' argument, an argument about what sort of political morality people would choose were they setting up society from an 'original position'. As Rawls says of the argument we have just looked at:

none of the preceding remarks [about equality of opportunity] are an argument for this conception [of justice], since in a contract theory all arguments, strictly speaking, are to be made in terms of what it would be rational to choose in the original position. But I am concerned here to prepare the way for the favored interpretation of the two principles of justice so that these criteria, especially the [difference principle], will not strike the reader as too eccentric or bizarre. (1971: 75)

So Rawls conceives his first intuitive argument as simply preparing the ground for the real argument, which is based on the idea of a social contract. This is an unusual strategy, for social contract arguments are usually thought of as being weak, and Rawls seems to be relegating a fairly strong argument into a back-up role behind the weaker social contract argument.

Why are social contract arguments thought to be weak? Because they seem to rely on very implausible assumptions. They ask us to imagine a state of nature before there is any political authority. Each person is on their own, in the sense that there is no higher authority with the power to command their obedience, or with the responsibility for protecting their interests or possessions. The question is, what kind of contract would such individuals, in the state of nature, agree to concerning the establishing of a political authority which would have these powers and responsibilities? Once we know what the terms of the contract are, we know what the government is obligated to do, and what the citizens are obliged to obey.

Different theorists have used this technique—Hobbes, Locke, Kant, Rousseau—and come up with different answers. But they have all been subject to the same criticism—namely, there never was such a state of nature, or such a contract. Hence neither citizens nor government are bound by it. Contracts only create obligations if they are actually agreed to. We can say that a certain agreement is the contract that people would have signed in some state of nature, and so is a hypothetical contract. But as Dworkin says, 'a hypothetical agreement is not simply a pale form of an actual contract; it is no contract at all' (Dworkin 1977: 151). The idea that we are bound by the contract we would accept in a state of nature implies that because a man would have consented to certain principles if asked in advance, it is fair to apply those principles to him later, under different circumstances, when he does not consent. But that is a bad argument. Suppose I did not know the value of my painting on Monday; if you had offered me $100 for it then I would have accepted. On Tuesday I discovered it was valuable. You cannot argue that it would be fair for the courts to make me sell it to you for $100 on Wednesday. It may be my good fortune that you did not ask me on Monday, but that does not justify coercion against me later. (Dworkin 1977: 152)

Thus the idea of a social contract seems either historically absurd (if it is based on actual agreement) or morally insignificant (if it is based on hypothetical agreement).

But, as Dworkin notes, there is another way to interpret social contract arguments. We should think of the contract not primarily as an agreement, actual or hypothetical, but as a device for teasing out the implications of certain moral premises concerning people's moral equality. We invoke the idea of a state of nature not to work out the historical origins of society, or the historical obligations of governments and individuals, but to model the idea of the moral equality of individuals.

Part of the idea of being moral equals is the claim that none of us is inherently subordinate to the will of others, none of us comes into the world as the property of another, or as their subject. We are all born free and equal. Throughout most of history, many groups have been denied this equality—in feudal societies, for example, peasants were viewed as naturally subordinate to aristocrats. It was the historical mission of classical liberals like Locke to deny this feudal premiss. And the way that they made clear their denial that some people were naturally subordinate to others was to imagine a state of nature in which people were equal in status. As Rousseau said, 'man is born free, and yet everywhere is in chains'. The idea of a state of nature does not, therefore, represent an anthropological claim about the pre-social existence of human beings, but a moral claim about the absence of natural subordination amongst human beings.

Classical liberals were not anarchists, however, who believe that governments are never acceptable. Anarchists believe that people can never come to have legitimate authority over others, and that people can never be legitimately compelled to obey such authority. Since these liberals were not anarchists, the pressing question was to explain how people born free and equal can come to be governed. Their answer was roughly this: due to the uncertainties and scarcities of social life, individuals, without giving up their moral equality, would endorse ceding certain powers to the state, but only if the state used these powers in trust to protect individuals from those uncertainties and scarcities. If the government betrayed that trust and abused its powers, then the citizens were no longer under an obligation to obey, and indeed had the right to rebel. Having some people with the power to govern others is compatible with respecting moral equality because the rulers only hold this power in trust, to protect and promote the interests of the governed.

This is the kind of theory that Rawls is adapting. As he puts it, 'my aim is to present a conception of justice which generalizes and carries to a higher level of abstraction the familiar theory of the social contract as found, say, in Locke, Rousseau, and Kant' (1971: 11). The point of the contract is to determine principles of justice from a position of equality: in Rawls's theory,

The original position of equality corresponds to the state of nature in the traditional theory of the social contract. The original position is not, of course, thought of as an actual historical state of affairs, much less as a primitive condition of culture. It is understood as a purely hypothetical situation characterized so as to lead to a certain conception of justice. (1971: 12)

While Rawls's original position 'corresponds' to the idea of a state of nature, it also differs from it, for Rawls believes that the usual state of nature is not really an 'initial position of equality' (1971: 11). This is where the contract argument joins up with his intuitive argument. The usual account of the state of nature is unfair because some people have more bargaining power than others—more natural talents, initial resources, or sheer physical strength—and they are able to hold out longer for a better deal, while those who are less strong or talented have to make concessions. The uncertainties of nature affect everyone, but some people can deal better with them, and they will not agree to a social contract unless it entrenches their natural advantages. This, we know, is unfair in Rawls's eyes. Since these natural advantages are undeserved, they should not privilege or disadvantage people in determining principles of justice.[2]

So a new device is needed to tease out the implications of moral equality, a device that prevents people from exploiting their arbitrary advantages in the selection of principles of justice. This is why Rawls develops the otherwise peculiar construction known as the 'original position'. In this revised original position, people are behind a 'veil of ignorance' so that

no one knows his place in society, his class position or social status, nor does any one know his fortune in the distribution of natural assets and abilities, his intelligence, strength, and the like. I shall even assume that the parties do not know their conceptions of the good or their special psychological propensities. The principles of justice are chosen behind a veil of ignorance. This ensures that no one is advantaged or disadvantaged in the choice of principles by the outcome of natural chance or the contingency of social circumstances. Since all are similarly situated and no one is able to design principles to favour his particular condition, the principles of justice are the result of a fair agreement or bargain. (1971: 12)

Many critics have viewed this demand that people distance themselves from knowledge of their social background and individual desires as evidence of a bizarre theory of personal identity. What is left of one's self when all that knowledge is excluded? It is difficult to imagine oneself behind such a veil of ignorance, much more difficult than imagining oneself in the traditional state of nature, where the fictional people were at least relatively whole in mind and body.

But the veil of ignorance is not an expression of a theory of personal identity. It is an intuitive test of fairness, in the same way that we try to ensure a fair division of cake by making sure that the person who cuts it does not know which piece she will get.[3] The veil of ignorance similarly ensures that those who might be able to influence the selection process in their favour, due to their better position, are unable to do so. As Rawls says,

One should not be misled, then, by the somewhat unusual conditions which characterize the original position. The idea here is simply to make vivid to ourselves the restrictions that it seems reasonable to impose on arguments for principles of justice, and therefore on these principles themselves. Thus it seems reasonable and generally accepted that no one should be advantaged or disadvantaged by natural fortune or social circumstances in the choice of principles. It also seems widely agreed that it should be impossible to tailor principles to the circumstances of one's own case. . . . In this manner the veil of ignorance is arrived at in a natural way. (1971: 18–19)

The original position is intended 'to represent equality between human beings as moral persons', and the resulting principles of justice are those which people 'would consent to as equals when none are known to be advantaged by social and natural contingencies'. We should look at the original position as 'an expository device' which 'sums up the meaning' of our notions of fairness and 'helps us to extract their consequences' (1971: 19, 21).

Rawls's argument is not, then, that a certain conception of equality is derived from the idea of a hypothetical contract. That would be subject to all the criticisms that Dworkin mentions. Rather, the hypothetical contract is a way of embodying a certain conception of equality, and a way of extracting the consequences of that conception for the just regulation of social institutions. By removing sources of bias and requiring unanimity, Rawls hopes to find a solution that is acceptable to everyone from a position of equality—i.e. that respects each person's claim to be treated as a free and equal being.

Since the premiss of the argument is equality, not contract, to criticize it we need to show that it fails to embody an adequate account of equality. It is not enough—indeed, it is irrelevant—to say that the contract is historically inaccurate, or that the veil of ignorance is psychologically impossible, or that the original position is in some other way unrealistic. The question is not whether the original position could ever really exist, but whether the principles which would be chosen in it are likely to be fair, given the nature of the selection process.

Even if we accept Rawls's idea of the social contract as a device for embodying a conception of equality, it is far from clear what principles would actually be chosen in the original position. Rawls, of course, thinks that the difference principle would be chosen. But his argument here is supposed to be independent of the first intuitive argument concerning equality of opportunity. As we have seen, he does not consider that kind of argument to be relevant, 'strictly speaking', within a contract theory. So the difference principle is just one of many possible choices which parties in the original position could make.

How do the principles of justice get chosen? The basic idea is this: while we do not know what position we will occupy in society, or what goals we will have, there are certain things we will want or need to enable us to lead a good life. Whatever the differences between individuals' plans of life, they all share one thing—they all involve *leading a life*. As Waldron puts it, 'there is something like *pursuing a conception of the good life* that all people, even those with the most diverse commitments, can be said to be engaged in . . . although people do not share one another's ideals, they can at least abstract from their experience a sense of *what it is like to be committed to an ideal of the good life*' (Waldron 1987: 145; cf. Rawls 1971: 92–5, 407–16). We are all committed to an ideal of the good life, and certain things are needed in order to pursue these commitments, whatever their more particular content. In Rawls's theory, these things are called 'primary goods'. There are two kinds of primary goods:

1. social primary goods—goods that are directly distributed by social institutions, like income and wealth, opportunities and powers, rights and liberties;
2. natural primary goods—goods like health, intelligence, vigour, imagination, and natural talents, which are affected by social institutions, but are not directly distributed by them.

In choosing principles of justice, people behind the veil of ignorance seek to ensure that they will have the best possible access to those primary goods distributed by social institutions (i.e. the social primary goods). This does not mean that egoism underlies our sense of justice. Since no one knows what position they will occupy, asking people to decide what is best for themselves has the same consequence as asking them to decide what is best for everyone considered impartially. In order to decide from behind a veil of ignorance which principles will promote my good, I must put myself in the shoes of every person in society and see what promotes their good, since I may end up being any one of those people. When combined with the veil of ignorance, therefore, the assumption of rational self-interest 'achieves the same purpose as benevolence' (Rawls 1971: 148), for I must sympathetically identify with every person in society and take their good into account as if it were my own. In this way, agreements made in the original position give equal consideration to each person.

So the parties in the original position are trying to ensure the best possible access to the primary goods that enable them to lead a worthwhile life, wiithout knowing where they will end up in society. There are still many different principles they could choose. They might choose an equal distribution of social primary goods for all social positions. But Rawls says that this is irrational when certain kinds of inequalities—e.g. those sponsored by the difference principle—improve everyone's access to primary goods. They might choose a utilitarian principle that instructs social institutions to distribute primary goods in such a way as to maximize utility in society. This would maximize the average utility that parties in the original position could expect to have in the real world, and, on some accounts of rationality, that makes it a rational choice. But it also involves the risk that you will be one of those who is endlessly sacrificed for the greater good of others. It leaves your liberties, possessions, and even your life vulnerable to the selfish and illegitimate preferences of others. Indeed, it leaves you unprotected precisely in those situations where you are most likely to need protection—e.g when your beliefs, skin-colour, sex, or natural abilities make you unpopular, or simply dispensable, to the majority. This makes utilitarianism an irrational choice, on some accounts of rationality, for it is rational to ensure your basic rights and resources are protected, even if you thereby lessen your chance of receiving benefits above and beyond the basic goods that you seek to protect.

So there are conflicting accounts of what it is rational to do in such a situation—the rationality of gambling versus the rationality of playing it safe. If we knew what the odds were of having our basic rights violated in a utilitarian society, then we would have a better idea of how rational it is to take the gamble. But the veil of ignorance excludes that information. The rationality of gambling also depends on whether one is personally risk-averse or not— some people do not mind taking risks, others prefer security. But the veil of ignorance excludes knowledge of personal tastes as well. What then is the rational choice? Rawls says that it is rational to adopt a 'maximin' strategy—that is, you *maximize* what you would get if you wound up in the *minimum*, or worst-off, position. As Rawls says, this is like proceeding on the assumption that your worst enemy will decide what place in society you will occupy (Rawls 1971: 152–3). As a result, you select a scheme that maximizes the minimum share allocated under the scheme.

For example, imagine that the following are the possible distributive schemes in a three-person world:

1. 10 : 8 : 1
2. 7 : 6 : 2
3. 5 : 4 : 4.

Rawls's strategy tells you to pick (3). If you do not know how likely it is that you will be in the best or worst position, the rational choice according to Rawls is the third scheme. For even if you end up in the worst position, it gives you more than you would get if you were in the bottom of the other schemes.

Notice that you should pick the third scheme even though the first two schemes have a higher average utility. The problem with the first two schemes is that there is some chance, unknown in size, that your life will be completely unsatisfactory. And since each of us has only one life to lead, it is irrational to accept the chance that your only life will be so unsatisfactory. So, Rawls concludes, people in the original position would select the difference principle. And this result happily matches what the first intuitive argument told us. People using a fair decision-procedure for selecting principles of justice come up with the same principles that our intuitions tell us are fair.

Many people have criticized Rawls's claim that 'maximin' is the rational strategy. Some claim that it is equally rational, if not more rational, to gamble on utilitarianism. Others claim that it is impossible to assess the rationality of gambling without knowing something about the odds, or about one's risk-aversion. These critics allege that Rawls only comes up with the difference principle because he rigs the description of the veil of ignorance so as to yield it, or because he makes gratuitous psychological assumptions which he is not entitled to make (e.g. Hare 1975: 88–107; Barry 1973: ch. 9).

(a) The convergence of the two arguments

There is some truth in these criticisms, but it is a misguided line of criticism. For Rawls admits that he rigs the description of the original position to yield the difference principle. He recognizes that 'for each traditional conception of justice there exists an interpretation of the initial situation in which its principles are the preferred solution', and that some interpretations will lead to utilitarianism (1971: 121). There are many descriptions of the original position that are compatible with the goal of creating a fair decision-procedure, and the difference principle would not be chosen in all of them. So before we can determine which principles would be chosen in the original position, we need to know which description of the original position to accept. And, Rawls says, one of the grounds on which we choose a description of the original position is that it yields the principles we find intuitively acceptable.

Thus, after saying that the original position should model the idea that people are moral equals, Rawls goes on to say that 'there is, however, another side to justifying a particular description of the original position. This is to see if the principles which would be chosen match our considered convictions of justice or extend them in an acceptable way' (1971: 19). Hence, in deciding on the preferred description of the original position we 'work from both ends'. If the principles chosen in one version do not match our convictions of justice, then

we have a choice. We can either modify the account of the initial situation or we can revise our existing judgements, for even the judgements we take provisionally as fixed points are liable to revision. By going back and forth, sometimes altering the conditions of the contractual circumstances, at others withdrawing our judgements and conforming them to principle, I assume that eventually we shall find a description of the initial situation that both expresses reasonable conditions and yields principles which match our considered judgements duly pruned and adjusted. (1971: 20)

So the intuitive argument and the contract argument are not independent after all. Rawls admits to modifying the original position in order to make sure that it yields principles which match our intuitions (at least those intuitions that we continue to hold after having engaged in this two-way process of harmonizing theory and intuitions). This may sound like cheating. But it only appears so if we take Rawls to be claiming that the two arguments provide entirely independent support for one another. And while he sometimes makes that claim, in other places he admits that the two arguments are interdependent, both drawing on the same set of considered intuitions.

But then why bother with the contract device? Why not just use the first intuitive argument? This is a good question. While the contract argument is not as bad as critics suggest, it is also not as good as Rawls suggests. If each theory of justice has its own account of the contracting situation, then we have to decide beforehand which theory of justice we accept, in order to know which description of the original position is suitable. Rawls's opposition to gambling away one life for the benefit of others, or to penalizing those with undeserved natural handicaps, leads him to describe the original position in one way; those who disagree with him on these issues will describe it another way. This dispute cannot be resolved by appeal to contractual agreement. It would beg the question for either side to invoke its account of the contracting situation in defence of its theory of justice, since the contracting situation presupposes the theory. All the major issues of justice, therefore, have to be decided beforehand, in order to know which description of the original position to accept. But then the contract is redundant.

This is not to say that the contract device is entirely useless. First, the original position provides a way to render vivid our intuitions, in the same way that earlier theorists invoked the state of nature to render vivid the idea of natural equality. Secondly, while the intuitions appealed to in the equal opportunity argument show that fair equality of opportunity is not enough, they do not tell us what more is required, and the contract device may help us render our intuitions more precise. This is what Rawls means by saying the device can help 'extract the consequences' of our intuitions. Thirdly, it provides a perspective from which we can test opposing intuitions. Someone who is naturally talented might sincerely object to the idea that talents are arbitrary. We would then have a clash of intuitions. But if that same person would no longer object were she ignorant of where she was going to end up in the natural lottery, then we can say with some confidence that our intuition was the right one, and that her opposing intuition was the result of opposing personal interests. Certain intuitions might seem less compelling when they are viewed from a perspective detached from one's own position in society. The contract argument tests our intuitions by showing whether they would be chosen from an unbiased perspective. The contract thus renders vivid certain general intuitions, and provides an impartial perspective from which we can consider more specific intuitions (Rawls 1971: 21–2, 586).

So there are benefits in employing the contract device. On the other hand, the contract device is not required for these purposes. As we saw in the last chapter, some theorists (e.g. Hare) invoke 'ideal sympathizers', rather than impartial contractors, to express the idea of equal consideration (ch. 2, s. 5b above). Both theories instruct the moral agent to adopt the impartial point of view, but whereas impartial contractors view each person in society as one of the possible future locations of their own good, ideal sympathizers view each person in society as one of the components of their own good, since they sympathize with and so share each person's fate. The two theories use different devices, but the difference is relatively superficial, for the key move in each theory is to force agents to adopt a perspective which denies them knowledge of, or any ability to promote, their own particular good. Indeed, it is often difficult to distinguish impartial contractors from ideal sympathizers (Gauthier 1986: 237–8; Diggs 1981: 277; Barry 1989: 77, 196).[4]

Equal consideration can also be generated without any special devices at all, just by asking agents to give equal consideration to others notwithstanding their knowledge of, and ability to promote, their own good (e.g. Scanlon 1982; Barry 1989: 340–8). Indeed, there is a curious sort of perversity in using either the contractarian or ideal sympathizer device to express the idea of moral equality. The concept of a veil of ignorance attempts to render vivid the idea that other people matter in and of themselves, not simply as a component of our own good. But it does so by imposing a perspective from which the good of others is simply a component of our own (actual or possible) good. The idea that people are ends in themselves gets obscured when we invoke 'the idea of a choice which advances the interests of a single rational individual for whom the various individual lives in a society are just so many different possibilities' (Scanlon 1982: 127; cf. Barry 1989: 214–15, 336, 370). Rawls tries to downplay the extent to which people in the original position view the various individual lives in society as just so many possible outcomes of a self-interested choice, but the contract device encourages that view, and so obscures the true meaning of equal concern.

So the contract device adds little to Rawls's theory. The intuitive argument is the primary argument, whatever Rawls says to the contrary, and the contract argument (at best) just helps express it. But it is not clear that Rawls needs an independent contract argument. Rawls had initially complained about the way people were forced to choose between utilitarianism, a systematic but counter-intuitive theory, and intuitionism, a collection of miscellaneous intuitions with no theoretical structure. If he has found a systematic alternative to utilitarianism which is in harmony with our intuitions, then his theory is a powerful one, in no way weakened by the interdependence of the intuitive and contract arguments. As Rawls says, 'a conception of justice cannot be deduced from self-evident premises or conditions on principles; instead, its justification is a matter of the mutual support of many considerations, of everything fitting together into one coherent view' (1971: 21). He calls this 'reflective equilibrium', and that is his aim. His principles of justice are mutually supported by reflecting on the intuitions we appeal to in our everyday practices, and by reflecting on the nature of justice from an impartial perspective that is detached from our everyday positions. Because Rawls is seeking such a reflective equilibrium, criticisms like those of Hare and Barry are overstated. For even if they are right that the difference principle would not be chosen in the original position as Rawls describes it, he could redefine the original position so as to yield the difference principle. That sounds like cheating, but it is useful and legitimate if in fact it leads us to a reflective equilibrium—if it means that 'we have done what we can to render coherent and to justify our convictions of social justice' (1971: 21).

A really successful criticism of Rawls must either challenge his fundamental intuitions, or show why the difference principle is not the best spelling out of these intuitions (and hence why a different description of the original position should be part of our reflective equilibrium). I will look at theories which challenge the basic intuitions in later chapters. But first I want to look at this second option. Can we find any problems internal to Rawls's theory, criticisms not of his intuitions, but of the way he develops them?

(b) Internal problems

One of Rawls's central intuitions, we have seen, concerns the distinction between choices and circumstances. His argument against the prevailing view of equality of opportunity depends heavily on the claim that it gives too much room for the influence of our undeserved natural endowments. I agreed with Rawls here. But Rawls himself leaves too much room for the influence of natural inequalities, and at the same time leaves too little room for the influence of our choices.

(i) Compensating for natural inequalities

I will look at the question of natural talents first. Rawls says that people's claim to social goods should not be dependent on their natural endowments. The talented do not deserve any greater income, and they should only receive more income if it benefits the less well off. So, according to Rawls, the difference principle is the best principle for ensuring that natural assets do not have an unfair influence.

But Rawls's suggestion still allows too much room for people's fate to be influenced by arbitrary factors. This is because Rawls defines the worst off position entirely in terms of people's possession of social primary goods—i.e. rights, opportunities, wealth, etc. He does not look at people's possession of natural primary goods in determining who is worst off. Two people are equally well off for Rawls (in this context) if they have the same bundle of social primary goods, even though one person may be untalented, physically handicapped, mentally disabled, or suffering from poor health. Likewise, if someone has even a small advantage in social goods over others, then she is better off on Rawls's scale, even if the extra income is not enough to pay for extra costs she faces due to some natural disadvantage—e.g. the costs of medication for an illness, or of special equipment for some handicap.

But why should the benchmark for assessing the justice of social institutions be the prospects of the least well off in terms of social goods? This stipulation conflicts with both the intuitive and contract arguments. In the contract argument, the stipulation is unmotivated in terms of the rationality of the parties in the original position. If, as Rawls says, health is as important as money in being able to lead a successful life, and if the parties seek to find a social arrangement that guarantees them the greatest amount of primary goods in the worst possible outcome (the maximin reasoning), then why would they not treat lack of health and lack of money as equally cases of being less well off for the purposes of social distribution? Every person recognizes that she would be less well off if she suddenly became disabled, even if her bundle of social goods remained the same. Why would she not want society also to recognize her disadvantage?

The intuitive argument points in the same direction. Not only are natural primary goods as necessary as social goods for leading a good life, but people do not deserve their place in the distribution of natural assets, and so it is wrong for people to be privileged or disadvantaged because of that place. As we have seen, Rawls thinks this intuition leads to the difference principle, under which people only receive extra rewards for their talents if doing so is to the benefit of the less well off: 'we are led to the difference principle if we wish to set up the social system so that no one gains or loses from his arbitrary place in the distribution of natural assets or his initial position in society without giving or receiving compensating advantages in return' (1971: 102). But that is wrong, or at least misleading. We are only led to the difference principle if by 'gains or loses' we mean gains or loses in terms of social goods. The difference principle ensures that the well endowed do not get more social goods just because of their arbitrary place in the distribution of natural assets, and that the handicapped are not deprived of social goods just because of their place. But this does not entirely 'mitigate the effects of natural accident and social circumstance' (1971: 100). For the well endowed still get the natural good of their endowment, which the handicapped undeservedly lack. The difference principle may ensure that I have the same bundle of social goods as a handicapped person. But the handicapped person faces extra medical and transportation costs. She faces an undeserved burden in her ability to lead a satisfactory life, a burden caused by her circumstances, not her choices. The difference principle allows, rather than removes, that burden.[5]

Rawls seems not to have realized the full implications of his own argument against the prevailing view of equality of opportunity. The position he was criticizing is this: (1) Social inequalities are undeserved, and should be rectified or compensated, but natural inequalities can influence distribution in accordance with equality of opportunity. Rawls claims that natural and social inequalities are equally undeserved, so (1) is 'unstable'. Instead, he endorses: (2) Social inequalities should be compensated, and natural inequalities should not influence distribution. But if natural and social inequalities really are equally undeserved, then (2) is also unstable. We should instead endorse: (3) Natural and social inequalities should be compensated. According to Rawls, people born into a disadvantaged class or race not only should not be denied social benefits, but also have a claim to compensation because of that disadvantage. Why treat people born with natural handicaps any differently? Why should they not also have a claim to compensation for their disadvantage (e.g. subsidized medicine, transportation, job training, etc.), in addition to their claim to non-discrimination?

So there are both intuitive and contract reasons for recognizing natural handicaps as grounds for compensation, and for including natural primary goods in the index which determines who is in the least well-off position. There are difficulties in trying to compensate for natural inequalities, as I will discuss in section 4*b* below. It may be impossible to do what our intuitions tell us is most fair. But Rawls does not even recognize the desirability of trying to compensate such inequalities.

(ii) Subsidizing people's choices

The second problem concerns the flip side of that intuition. People do not deserve to bear the burden of unchosen costs, but how should we respond to people who choose to do costly things? We normally feel that unchosen costs have a greater claim on us than voluntarily chosen costs. We feel differently about someone who spends $100 a week on expensive medicine to control an unchosen illness, compared with someone who spends $100 a week on expensive wine because they enjoy its taste. Rawls appeals to this intuition when criticizing the prevailing view for being insensitive to the unchosen nature of natural inequalities. But how should we be sensitive to people's choices?

Imagine that we have succeeded in equalizing people's social and natural circumstances. To take the simplest case, imagine two people of equal natural talent who share the same social background. One wants to play tennis all day, and so only works long enough at a nearby farm to earn enough money to buy land for a tennis-court, and to sustain his desired lifestyle (i.e. food, clothing, equipment). The other person wants a similar amount of land to plant a garden, in order to produce and sell vegetables for herself and others. Furthermore, let us imagine, with Rawls, that we have started with an equal distribution of resources, which is enough for each person to get their desired land, and start their tennis and gardening. The gardener will quickly come to have more resources than the tennis-player, if we allow the market to work freely. While they began with equal shares of resources, he will rapidly use up his initial share, and his occasional farm work only brings in enough to sustain his tennis-playing. The gardener, however, uses her initial share in such a way as to generate a steadier and larger stream of income through larger amounts of work. Rawls would only allow this inequality if it benefits the least well off—i.e. if it benefits the tennis-player who now lacks much of an income. If the tennis-player does not benefit from the inequality, then the government should transfer some of her income to him in order to equalize income.

But there is something peculiar about saying that such a tax is needed to enforce equality, where that is understood to mean treating both people as equals. Remember that the tennis-player has the same talents as the gardener, the same social background, and started with an equal allotment of resources. As a result, he could have chosen income-producing gardening if he wished, just as she could have chosen non-income-producing tennis. They both faced a range of options which offered varying amounts and kinds of work, leisure, and income. Both chose that option which they preferred. The reason he did not choose gardening, therefore, is that he preferred playing tennis to earning money by gardening. People have different preferences about when it is worth giving up potential leisure to earn more income, and he preferred leisure while she preferred income.

Given that these differences in lifestyle are freely chosen, how is he treated unequally by allowing her to have the income and lifestyle that he did not want? Rawls defends the difference principle by saying that it counteracts the inequalities of natural and social contingencies. But these are not relevant here. Rather than removing a disadvantage, the difference principle simply makes her subsidize his expensive desire for leisure. She has to pay for the costs of her choice—i.e. she forgoes leisure in order to get more income. But he does not have to pay for the costs of his choice—i.e. he does not forgo income in order to get more leisure. He expects and Rawls requires that she pay for the costs of her own choices, and also subsidize his choice. That does not promote equality, it undermines it. He gets his preferred lifestyle (leisureful tennis), plus some income from her taxes, while she gets her preferred lifestyle (income-producing gardening) minus some income that is taxed from her. She has to give up part of what makes her life valuable in order that he can have more of what he finds valuable. They are treated unequally in this sense, for no legitimate reason.

When inequalities in income are the result of choices, not circumstances, the difference principle creates, rather than removes, unfairness. Treating people with equal concern requires that people pay for the costs of their own choices. Paying for choices is the flip side of our intuition about not paying for unequal circumstances. It is unjust if people are disadvantaged by inequalities in their circumstances, but it is equally unjust for me to demand that someone else pay for the costs of my choices. In more technical language, a distributive scheme should be 'endowment-insensitive' and 'ambition-sensitive' (Dworkin 1981: 311). People's fate should depend on their ambitions (in the broad sense of goals and projects about life), but should not depend on their natural and social endowments (the circumstances in which they pursue their ambitions).

Rawls himself emphasizes that we are responsible for the costs of our choices. This in fact is why his account of justice measures people's share of primary goods, not their level of welfare. Those who have expensive desires will get less welfare from an equal bundle of primary goods than those with more modest tastes. But, Rawls says, it does not follow that those with modest tastes should subsidize the extravagant, for we have 'a capacity to assume responsibility for our ends'. Hence 'those with less expensive tastes have presumably adjusted their likes and dislikes over the course of their lives to the income and wealth they could reasonably expect; and it is regarded as unfair that they now should have less in order to spare others from the consequences' of their extravagance (Rawls 1982*b*: 168–9; cf. 1975: 553; 1980: 545; 1974: 643; 1978: 63; 1985: 243–4). So Rawls does not wish to make the gardener subsidize the tennis-player. Indeed he often says that his conception of justice is concerned with regulating inequalities that affect people's life-chances, not the inequalities that arise from people's life-choices, which are the individual's own responsibility (1971: 7, 96; 1978: 56; 1979: 14–15; 1982*b*: 170). Unfortunately, the difference principle does not make any such distinction between chosen and unchosen inequalities. Hence one possible result of the difference principle is to make some people pay for others' choices, should it be the case that those with the least income are, like the tennis-player, in that position by choice. Rawls wants the difference principle to mitigate the unjust effects of natural and social disadvantage, but it also mitigates the legitimate effects of personal choice and effort.

So while Rawls appeals to this choices–circumstances distinction, his difference principle violates it in two important ways. It is supposed to mitigate the effect of one's place in the distribution of natural assets. But because Rawls excludes natural primary goods from the index which determines who is least well off, there is in fact no compensation for those who suffer undeserved natural disadvantages. Conversely, people are supposed to be responsible for the costs of their choices. But the difference principle requires that some people subsidize the costs of other people's choices. Can we do a better job of being 'ambition-sensitive' and 'endowment-insensitive'? This is the goal of Dworkin's theory.

4. DWORKIN ON EQUALITY OF RESOURCES

Dworkin accepts the 'ambition-sensitive' and 'endowment-insensitive' goal that motivated Rawls's difference principle. But he thinks that a different distributive scheme can do a better job living up to that ideal. His theory is a complicated one—involving the use of auctions, insurance schemes, free markets, and taxation—and it is impossible to lay out the whole theory. But I will present some of its central intuitive ideas.

(a) Paying for one's choices: the ambition-sensitive auction

I will start with Dworkin's account of an ambition-sensitive distributive scheme. For simplicity's sake, I will assume again that everyone has the same natural talents (I examine Dworkin's answer to the problem of unequal natural endowments later). Dworkin asks us to imagine that all of society's resources are up for sale in an auction, to which everyone is a participant. Everyone starts with an equal amount of purchasing power—100 clamshells, in his example—and people use their clamshells to bid for those resources that best suit their plan of life.

If the auction works out, everyone will be happy with the result, in the sense that they do not prefer anyone else's bundle of goods to their own. If they did prefer a different bundle, they could have bid for it, rather than the goods they did bid for. This generalizes the case of the tennis-player and gardener who, starting with the same amount of money, acquire the land they need for their desired activities. If the auction works, this will be true of everyone—i.e. each person will prefer their own bundle of goods to anyone else's. Dworkin calls this the 'envy test', and if it is met, then people are treated with equal consideration, for differences between them simply reflect their different ambitions, their different beliefs about what gives value to life. A successful auction meets the envy test, and makes each person pay for the costs of their own choices (Dworkin 1981: 285).

This idea of the envy test expresses the liberal egalitarian view of justice in its most defensible form. If it could be perfectly enforced, the three main aims of Rawls's theory would be fulfilled, i.e. respecting the moral equality of persons, mitigating the effects of morally arbitrary disadvantages, and accepting responsibility for our choices. Such a distributive scheme would be just, even though it allows some inequality in income. The gardener and tennis-player have unequal income, but there is no inequality in respect and concern, since each of them is able to lead the life they choose, each has an equal ability to bid for that bundle of social goods that best serves their beliefs about what gives value to life. To put it another way, no one can claim to be treated with less consideration than another in the distribution of resources, for if someone preferred another person's bundle of social goods, she could have bid for it instead. It is difficult to see how I could have a legitimate complaint against anyone else, or they against me.[6]

(b) Compensating natural disadvantages: the insurance scheme

Unfortunately, the auction will only meet the envy test if we assume that no one is disadvantaged in terms of natural assets. In the real world, the auction will fail the envy test, for some of the differences between people will not be chosen. Someone with handicaps or congenitally poor health may be able to bid for the same bundle of social goods as others, but she has special needs, and so her 100 clamshells will leave her less well off than others. She would prefer to be in their circumstances, without the handicap.

What should we do with natural disadvantages? Dworkin has a complex answer to that question, but we can prepare the way for it by looking at a simpler answer. The disadvantaged person faces extra burdens in leading a good life, burdens that cut into her 100 clamshells. Why not pay for all those extra costs before the auction, out of the general stock of social resources, and then divide up the remaining resources equally through the auction? Before the auction, we give the disadvantaged enough social goods to compensate for their unchosen inequality in natural assets. Once that is done, we give each person an equal share of the remaining resources to use in accordance with their choices in the auction. The auction results would now meet the envy test. Compensation before the auction would ensure that each person is equally able to choose and pursue a valuable life-plan; equal division of resources within the auction ensures that those choices are fairly treated. Hence the distribution would be both endowment-insensitive and ambition-sensitive.

This simple answer will not work. Extra money can compensate for some natural disadvantages—some physically handicapped people can be as mobile as able-bodied people if we provide the best technology available (which may be expensive). But that goal is impossible to achieve in other cases, for no amount of social goods will fully compensate for certain natural disadvantages. Imagine someone who is multiply handicapped and/or suffering from an incurable disease. Providing extra money can buy medical equipment, or assistance from skilled personnel, things which ensure there is no unnecessary pain in his life. And more money can always help a little more in terms of equipment or extending life. But none of this can ever put him in a situation where his circumstances are genuinely equalized. No amount of money can enable the severely disadvantaged person to lead as good a life as other people.

Full equality of circumstances is impossible. We could try to equalize circumstances as much as possible. But that too seems unacceptable. Since each additional bit of money might help the severely disadvantaged person, yet is never enough to equalize circumstances fully, we might be required to give all our resources to people with such handicaps, leaving nothing for everyone else (Dworkin 1981: 242, 300; cf. Fried 1978: 120–8). If resources had to be used to equalize circumstances first (before the auction starts), there would be none left for us to act on our choices (bidding for goods in the auction). But one of our goals in equalizing circumstances was precisely to allow each person to act on their chosen life-plans. Our circumstances affect our ability to pursue our ambitions. That is why they are morally important, why inequalities in them matter. Our concern for people's circumstances is a concern to promote their ability to pursue their ends. If in trying to equalize the means we prevent anyone from achieving their ends, then we have failed completely.

If we cannot achieve full equality of circumstances, and we should not always try to achieve it, what should we do? Given these difficulties, Rawls's refusal to compensate for natural disadvantages makes sense. Including natural disadvantages in the index which determines the least well off seems to create an insoluble problem. We do not want to ignore such disadvantages, but nor can we equalize them, and what could be in between, other than *ad hoc* acts of compassion or mercy?

Dworkin's proposal is similar to Rawls's idea of an original position. We are to imagine people behind a modified veil of ignorance. They do not know their place in the distribution of natural talents, and are to assume that they are equally susceptible to the various natural disadvantages which might arise. We give each person an equal share of resources—the 100 clamshells—and ask them how much of their share they are willing to spend on insurance against being handicapped, or otherwise disadvantaged, in the distribution of natural endowments. People might be willing to spend 30 per cent of their bundle of resources, for example, on such insurance, which would buy them a certain level of coverage for the different disadvantages they may suffer. If we can make sense of this hypothetical insurance market, and find a determinate answer to the question of what insurance people would buy in it, then we could use the tax system to duplicate the results. Income tax would be a way of collecting the premiums that people hypothetically agreed to pay, and the various welfare, medicare, and unemployment schemes would be ways of paying out the coverage to those who turned out to suffer from the natural disadvantages covered by the insurance.

This provides the middle ground between ignoring unequal natural assets and trying in vain to equalize circumstances. It would not lead to ignoring the problem, for everyone would buy some insurance. It is irrational not to provide any protection against the calamities that may befall you. But no one would spend all of their clamshells on insurance, since they would have nothing left to pursue their goals with. The amount of society's resources that we dedicate to compensating for natural disadvantages is limited to the coverage people would buy through premiums paid out of their initial bundle (Dworkin 1981: 296–9).

Some people are still disadvantaged in undeserved ways under this scheme, so we have not found the pure ambition-sensitive and endowment-insensitive distribution we were looking for. But we cannot achieve that goal no matter what we do, so we need a theory of the 'second-best'. Dworkin claims that his insurance scheme is fair as a second-best theory, because it is the result of a decision-procedure which is fair. It is generated by a procedure which treats everyone as equals, and excludes obvious sources of unfairness, so that no one is in a privileged position in buying the insurance. Hopefully everyone can recognize and accept the fairness of letting their compensation be determined by what they would have chosen in such a hypothetical position of equality.

It might seem that Dworkin's unwillingness to try as best we can to mitigate the effects of natural disadvantages shows an in-adequate regard for the well-being of the disadvantaged. After all, they did not choose to be disadvantaged. But if we attempted to provide the highest possible coverage to those who turn out to be disadvantaged, the result would be the 'slavery of the talented'. Consider the situation of those who must pay for the insurance without receiving any compensation:

Someone who 'loses' in this sense must work hard enough to cover his premium before he is free to make the tradeoffs between work and consumption he would have been free to make if he had not insured. If the level of coverage is high then this will enslave the insured, not simply because the premium is high, but because it is extremely unlikely that his talents will much surpass the level that he has chosen, which means that he must indeed work at full stretch, and that he will not have much choice about what kind of work to do. (Dworkin 1981: 322)

Those who were fortunate in the natural lottery would be forced to be as productive as possible in order to pay the high premiums they had hypothetically bought against natural disadvantage. The insurance scheme would cease to be a constraint that the talented could reasonably be expected to recognize in deciding how to lead their lives, but would rather become the determining factor in their lives. Their talents would be a liability that restricted their options, rather than a resource that expanded their options. The insurance scheme would have the effect that those with greater talents would have less freedom to choose their preferred leisure–consumption mix than those with lesser talents. Hence, equal concern for both the advantaged and the disadvantaged requires something other than maximal redistribution to the disadvantaged, even though it will leave the disadvantaged envying the well endowed.[7]

Jan Narveson says that this failure to ensure a real-world fulfilment of the envy test undermines Dworkin's theory. Suppose Smith is born with natural disadvantages relative to Jones, so that Jones is able to earn a larger income. Even if we tax Jones to fulfil the insurance obligations arising out of this hypothetical auction, Jones will still have more income than Smith, an undeserved inequality. As Narveson puts it, 'The *fact* is that Smith is, in every measure that matters to him or to Jones, way behind Jones in the actual world. Can we hold with a straight face that the bundle of counterfactuals added to his bundle of de facto resources sufficiently "compensates" him in the terms of a substantial theory of equality?' (Narveson 1983: 18). The envy test fails in the world, and, as Narveson says, it seems peculiar to say we have compensated for that by satisfying the envy test in some hypothetical situation.

But this objection begs the question. If we cannot fully equalize real-world circumstances, then what else can we do to live up to our convictions about the arbitrariness of one's place in the distribution of natural and social circumstances? Dworkin does not say that his scheme fully compensates for undeserved inequalities, just that it is the best we can do to live up to our convictions of justice. To criticize him, we need to show either how we can do better living up to those beliefs, or why we should not try to live up to them. Narveson does neither.

(c) The real-world equivalents: taxes and redistribution

Assuming that the insurance model is a legitimate, albeit second-best, response to the problem of equalizing circumstances, how can we apply it to the real world? It cannot be a matter of enforcing real insurance contracts, for the insurance market was purely hypothetical. So what in the world corresponds to the buying of premiums and the giving out of coverage benefits? I said earlier that we can use the tax system to collect premiums from the naturally advantaged, and use welfare schemes as a way of paying out the coverage to those who are disadvantaged. But the tax system can only approximate the results of the insurance scheme, for two reasons (Dworkin 1981: 312–14).

Firstly, there is no way of measuring, in the real world, what people's relative advantages and disadvantages are. One reason is that one of the things people choose to do with their lives is develop their talents. People who started with equal natural talents could come later to have differing skill levels. Those sorts of differences do not deserve compensation, since they reflect differences in choices. People who start off with greater skills may also develop them further, and then differences in talents will partially reflect different natural talents, and partially reflect different choices. In such cases, some but not all of the differences in talents deserve compensation. This will be extremely difficult to measure.

Moreover, it is impossible to determine in advance of the auction what counts as a natural advantage. That depends on what sorts of skills people value, which depends in turn on the goals they have in life. Certain skills (e.g. physical strength) are less important now than before, while others (e.g. abstract mathematical thought) are far more valuable. There is no way to know, in advance of people's choices, which natural capacities are advantages and which are disadvantages. This criterion changes continuously (if not radically), and it would be impossible to monitor these shifts.

How then can we fairly implement this insurance scheme, given the impossibility of identifying those rewards which accrue from talents rather than ambitions? Dworkin's answer is perhaps rather disappointing: we tax the rich, even though some got there purely by effort with no natural advantage, and support the poor, even though some, like the tennis-player, are there by choice without any natural disadvantage. Hence some people will get less coverage than they hypothetically bought, just because they are now, by dint of effort, in the upper income categories. And some people will get more coverage than they deserve, just because they have expensive lifestyles.

A second problem with applying the model is that natural disadvantages are not the only source of unequal circumstances (even in a society with equality of opportunity for different races, classes, or sexes). In the real world we lack full information, so that the envy test can be violated when unexpected things occur. A blight may ruin our gardener's crop for a number of years, leaving her with little income. But, unlike the tennis-player, she did not choose to lead an unproductive lifestyle. That was a wholly unforeseen natural contingency, and it would be wrong to make her pay for all the costs of her chosen lifestyle. If she had known it would be so costly, she would have chosen a different life-plan (unlike the tennis-player, who was aware of the costs of his lifestyle). These sorts of unexpected costs need to be fairly dealt with. But if we try to compensate for them through an insurance scheme similar to the one for natural talents, the result will have all the shortcomings of that other insurance scheme.

We now have two sources of deviation from the ideal of an ambition-sensitive, endowment-insensitive distribution. We want people's fate to be determined by the choices they make from a fair and equitable starting-point. But the idea of an equal starting-point includes not only an unachievable compensation for unequal endowment, but also an unachievable knowledge of future events. The former is needed to equalize circumstances, the latter is needed to know the costs of our choices, and hence be held responsible for them. The insurance scheme is a second-best response to these problems, and the taxation scheme is a second-best response to the problem of applying the insurance scheme. Given this distance between the ideal and the practice, it is inevitable that some people are undeservedly penalized for their unfortunate circumstances, while others are undeservedly subsidized for the costs of their choices.

Can we not do any better in achieving an ambition-sensitive, endowment-insensitive distribution? Dworkin concedes that we could achieve one or other of the aims more completely. However, the two aims pull in opposite directions—the more we try to make the distribution sensitive to people's ambitions, the more likely it is that some people disadvantaged by circumstances will be undeservedly penalized, and vice versa. These are both deviations from the ideal, and equally important deviations, so a proposal which concentrates on one to the exclusion of the other is unacceptable. We must employ both criteria, even if the effect is that neither is fully satisfied (Dworkin 1981: 327–8, 333–4).

This is a rather disappointing conclusion. Dworkin argues persuasively that a just distribution must identify 'which aspects of any persons's economic position flow from his choices and which from advantages and disadvantages that were not matters of choice' (Dworkin 1985: 208). But it seems that in practice his ideal is 'indistinguishable in its strategic implications' from theories, like Rawls's difference principle, which do not mark this distinction (Carens 1985: 67; cf. Dworkin 1981: 338–44). The hypothetical calculations Dworkin's theory requires are so complex, and their institutional implementation so difficult, that its theoretical advantages cannot be translated into practice (Mapel 1989: 39–56; Carens 1985: 65–7; cf. Varian 1985: 115–19; Roemer 1985a).

None the less, Dworkin's theory is important. His idea of the envy test describes, and makes vivid, what it would be for a distribution scheme to fulfil the basic aims of Rawls's theory: a distributive scheme that respects the moral equality of persons by compensating for unequal circumstances while holding individuals responsible for their choices. There may be a more appropriate apparatus for implementing these ideas than the mixture of auctions, insurance schemes, and taxes that Dworkin employs. But if we accept these fundamental premisses, Dworkin has helped us clarify their consequences for distributive justice.

It is worth pausing for a moment and reviewing the arguments presented so far. I started by examining utilitarianism, which is attractive for its insistence on interpreting morality in terms of a concern for the welfare of human beings. But that concern, which we saw was an egalitarian one, need not require the maximization of welfare. The utilitarian idea of giving equal weight to each person's preferences has some initial plausibility as a way of showing equal concern for people's welfare. But on inspection, utilitarianism often violates our sense of what it is to treat people as equals, especially in its lack of a theory of fair shares. This was Rawls's motivation for developing a conception of justice that provides a systematic alternative to utilitarianism. When we examined prevailing ideas about fair shares, we encountered the belief that it is unfair for people to be penalized for matters of brute luck, for circumstances which are morally arbitrary and beyond their control. This is why we demand equal opportunity for people from different racial and class backgrounds. But the same intuition should also tell us to recognize the arbitrary nature of people's place in the distribution of natural assets. This is the motivation for Rawls's difference principle, under which the more fortunate only receive extra resources if it benefits the unfortunate.

But the difference principle is both an over-reaction and also an insufficient reaction to the problem of undeserved inequalities. It is insufficient in not providing any compensation for natural disadvantages; and it is an over-reaction in precluding inequalities that reflect different choices, rather than different circumstances. We want a theory to be more ambition-sensitive and less endowment-sensitive than Rawls's difference principle. Dworkin's theory aspires to these twin goals. But we saw that these goals are unreachable in their pure form. Any theory of fair shares will have to be a theory of the second-best. Dworkin's scheme of auctions and insurance is one suggestion for fairly resolving the tension between these two core goals of the liberal conception of equality.

So Dworkin's theory was a response to problems in Rawls's conception of equality, just as Rawls's theory was a response to problems in the utilitarian conception of equality. Each can be seen as attempting to refine, rather than reject, the basic intuitions which motivated the previous one. Rawls's egalitarianism is a reaction against utilitarianism, but is also partly a development from utilitarianism's core intuitions, and the same is true of Dworkin's relation to Rawls. Each theory defends its own principles by appealing to the very intuitions that led people to adopt the previous theory.

5. THE POLITICS OF LIBERAL EQUALITY

One common way of describing the political landscape is to say that liberals seek a compromise between libertarians on the right, who believe in liberty, and Marxists on the left, who believe in equality. This is supposed to explain why liberals endorse the welfare state, which combines capitalist freedoms and inequalities with various egalitarian welfare policies (e.g. Sterba 1988: 31). But that description is inaccurate, at least for the liberals I have looked at. If they allow some kinds of inequality-producing economic freedoms, it is not because they believe in liberty as opposed to equality. Rather, they believe that such economic freedoms are needed to enforce their more general idea of equality itself. The same principle that tells liberals to allow market freedom—i.e. that it holds people responsible for their choices—also tells them to limit the market where it penalizes people for reasons other than their choices. The same conception of equality underlies both market freedom and its constraint. Hence the liberal favours a mixed economy and welfare state not in order to compromise conflicting ideals, 'but to achieve the best practical realization of the demands of equality itself' (Dworkin 1978: 133; 1981: 313, 338).

But would the implementation of this theory actually lead to our familiar welfare state? It is difficult to say exactly what policies the theory requires. It suggests a familiar mixture of market freedoms and state taxation. Yet it also demands that each person start their life with an equal share of society's resources, which is a striking attack on the entrenched divisions of class, race, and gender in our society. Quite radical government policies might be required to eliminate those entrenched hierarchies—e.g. nationalizing wealth, affirmative action, worker self-ownership, payment to home-makers, public health care, free university education, etc. We would have to look at these one by one to see if they move us closer to the results of the hypothetical auction, and that answer will often depend on the particular circumstances. Perhaps liberal equality would favour something like our existing schemes for ongoing income redistribution, but only after a radical one-time redistribution of wealth and property-ownership (Krouse and McPherson 1988: 103). Dworkin's theory does not answer these questions, it just provides the framework in which those discussions can take place.

While it is difficult to know exactly what Dworkin's theory will mean in practice, it seems certain that liberalism's institutional commitments have not kept pace with its theoretical commitments. William Connolly says that liberalism's theoretical premises can be united with its traditional institutions 'as long as it is possible to believe that the welfare state in the privately incorporated economy of growth can be the vehicle of liberty and justice' (Connolly 1984: 233). He claims, however, that the demands of the private economy conflict with the principles of justice that underlie the welfare state. The welfare state needs a growing economy to support its redistributive programmes, but the structure of the economy is such that growth can only be secured by policies inconsistent with the principles of justice that underlie those welfare programmes (Connolly 1984: 227–31).

According to Connolly, this has led to a 'bifurcation of liberalism'. One stream clings to the traditional institutions of liberal practice, and exhorts people to lower their expectations concerning justice and freedom. The other stream (in which he includes Dworkin) reaffirms the principles, but 'the commitment to liberal principles is increasingly matched by the disengagement from practical issues . . . this principled liberalism is neither at home in the civilization of productivity nor prepared to challenge its hegemony' (Connolly 1984: 234). I think this accurately describes the condition of contemporary liberalism. The ideals of liberal equality are compelling, but they require reforms that are more extensive than Rawls or Dworkin have explicitly allowed. Neither has challenged the 'civilization of productivity' whose maintenance has involved the perpetuation and often exaggeration of entrenched inequalities of race, class, and gender.

Dworkin often writes as if the most obvious or likely result of implementing his conception of justice would be to increase the level of transfer payments between occupants of existing social roles (e.g. Dworkin 1981: 321; 1985: 208). But his theory entails a more radical reform—namely, a change in the way existing roles are defined. As he recognizes, important components of the resources available to a person include opportunities for skill-development, personal accomplishment, and the exercise of responsibility. These are predominantly matters, not of the material rewards for a given job, but of the social relations entailed by the job. People would not generally choose to enter social relations that deny these opportunities, or that put them in relations of domination or degradation. From a position of equality, women would not have agreed to a system of social roles that defines 'male' jobs as superior to, and dominating of, 'female' jobs. And workers would not have agreed to the exaggerated distinction between 'mental' and 'manual' labour. We know that people in a position of initial equality would not have chosen these roles, for they were created without the consent of women and workers, and in fact often required their legal and political suppression. For example, the division of authority between doctors and nurses was opposed by women health care practitioners (Ehrenreich and English 1973: 19–39), and the 'scientific management' system was opposed by workers (Braverman 1974). Both changes would have taken a substantially different form if women and workers had had the same power as men and capitalists.

Dworkin says that increased transfer payments are justified

because we can assume that the poor would be willing to do the
work in higher-paying jobs if they entered the market on an equal
footing (Dworkin 1985: 207). But we can also assume that if the
poor entered the market on an equal footing, they would not accept
relations of unequal power and domination. We have as good
evidence for the latter as for the former. Liberals, therefore, should
not only redistribute income from doctors to nurses, or from
capitalists to workers, but should also ensure that doctors and
capitalists do not have the power to define relationships of
dominance. Justice requires that people's situation match the
results of the hypothetical tests that Rawls and Dworkin employ,
not only in terms of income, but also in terms of social power.
Focusing solely on income redistribution is to make 'the great error
of reformers and philanthropists . . . to nibble at the consequences
of unjust power, instead of redressing the injustice itself' (Mill
1965: 953).

It is interesting to note that Rawls himself denies that the
principles of liberal equality can be met by the welfare state. He
endorses the quite different idea of a 'property-owning democracy'
(1971: 274). The difference has been described this way:

welfare state capitalism (as commonly understood) accepts severe class
inequality in the distribution of physical and human capital, and seeks to
reduce the consequent disparities in market outcomes through redistributive tax and transfer programmes. Property-owning democracy, by
contrast, aims at sharply reduced inequality in the underlying distribution
of property and wealth, and greater equality of opportunity to invest in
human capital, so that the operation of the market generates smaller
inequalities to begin with. Thus, the two alternative regimes exemplify two
alternative strategies for providing justice in political economy: Welfare-
state capitalism accepts as given substantial inequality in the initial
distribution of property and skill endowments, and then seeks to
redistribute income ex post; property-owning democracy seeks greater
equality in the ex ante distribution of property and skill endowments, with
correspondingly less emphasis upon subsequent redistributive measures.
(Krouse and McPherson 1988: 84)[8]

Attacking inequality in this way, Rawls says, will prevent relations
of domination and degradation within the division of labour: 'no
one need be servilely dependent on others and made to choose
between monotonous and routine occupations which are deadening
to human thought and sensibility' (Rawls 1971: 529, 281; cf.
Krouse and McPherson 1988: 91–2; DiQuattro 1983: 62–3).

Unfortunately, Rawls does not provide much of a description of this property-owning democracy—as one critic puts it, 'these points never find their way into the substance of his theory of justice' (Doppelt 1981: 276). Other than a rather modest proposal to limit inheritances, Rawls gives us no idea how to confront the injustices in our society, or indeed whether he thinks there are substantial injustices to be confronted. Hence it is understandable that most critics view Rawls as offering 'a philosophical apologia for an egalitarian brand of welfare-state capitalism' (Wolff 1977: 195; cf. Doppelt 1981: 262; Clark and Gintis 1978: 311–14).

According to Dworkin, the egalitarian premiss underlying Rawls's (and his own) theory 'cannot be denied in the name of any more radical concept of equality, because none exists' (Dworkin 1977: 182). In fact, it seems that that premiss has more radical implications than either Dworkin or Rawls recognizes, implications that traditional liberal institutions are unable to accommodate. It might be that a full implementation of Rawlsian or Dworkinian justice would move us closer to market socialism than to welfare-state capitalism (Buchanan 1982: 124–31, 150–2; DiQuattro 1983). It might also move us closer to radical changes in gender relations. The welfare state has not stopped the increasing feminization of poverty, and if present trends continue, all of the people below the poverty line in America in the year 2000 will be women or children. Needless to say, such a maldistribution does not match the results of freely made choices in either Rawls's original position or Dworkin's auction. Yet neither theorist has anything to say about how this systematic devaluation of the roles of women can be removed. Indeed, Rawls defines his original position (as an assembly of 'heads of families'), and his principles of distribution (as measuring 'household income'), in such a way that questions about the justice of arrangements within the family are ruled out of court by definition (Okin 1987: 49). Of all the issues from which contemporary liberals have become disengaged, sexual inequality is the most glaring omission, and the one which liberal institutions seem least able to deal with (see ch. 7, s. 1 below).

So the relationship between contemporary liberal theory and traditional liberal political practice is unclear. The two have become disengaged in a number of ways. Liberalism is often called 'mainstream' political theory, as opposed to radical or critical theory. That label is accurate in one sense, for Rawls and Dworkin are trying to articulate and defend the ideals that they believe are at the moral base of our liberal-democratic culture. But it is inaccurate in another sense, if it implies that liberal theories are committed to defending all aspects of mainstream liberal politics, or to rejecting all aspects of the political programmes of other traditions. It would be wrong to assume that the account of liberal equality I have presented is necessarily tied to any particular liberal institution, or is necessarily opposed to any particular socialist or feminist proposal. We will have to wait until we examine these other theories before we can determine the extent of their differences with liberal equality.

Some people argue that if liberals endorse these more radical reforms, they have abandoned their liberalism. That seems unduly restrictive, given the historical ties between liberalism and radicalism (Gutmann 1980). It is also misleading, for however far liberal principles take us from traditional liberal practices, they are still distinctively liberal principles. I have argued in this section that liberals need to think seriously about adopting more radical politics.[9] In subsequent chapters, I will argue that radical theorists need to think seriously about adopting liberal principles. Just as liberal practice often does a disservice to liberal principles, I will argue that radical principles often do a disservice to radical politics. But first I will look at a theory which thinks liberals have gone too far in the direction of social and economic equality.

NOTES

1. Rawls has a number of subsidiary arguments for his two principles of justice. For example, Rawls says that his principles meet the requirements of 'publicity' (Rawls 1971: 133) and 'stability' (1971: 176–82) more fully than alternative accounts of justice. Principles of justice must be publicly known and easily applied, and the corresponding sense of justice must be stable and self-reinforcing (e.g. the 'strains of commitment' must not be too great). Rawls sometimes puts considerable weight on such arguments in defending his theory, but they do not by themselves generate a determinate theory of justice, and hence are subsidiary to the two main arguments I discuss. For a summary of the subsidiary arguments, see Parekh (1982: 161–2).
2. It is this condemnation of the unfairness inherent in the traditional state of nature which sets Rawls apart from another contract tradition—a

tradition which runs from Hobbes to recent theorists like David Gauthier and James Buchanan. Like Rawls, they hope to generate principles for regulating social life from the idea of an agreement in an initial position. But unlike Rawls, the agreement aims at mutual advantage, not justice, and so it is permissible, and indeed essential, that the initial situation reflect the differences in bargaining power that occur in the real world. I will discuss this second contractarian approach in chapter 4, and ask whether theories of mutual advantage should be considered as theories of justice at all.

3. Rawls says that the case of choosing principles of justice in the original position is importantly different from cutting a cake without knowing which piece you will get. He calls the first case an example of 'pure procedural justice', while the second is 'perfect procedural justice'. In each case, a procedure is supposed to yield just results. But in the former case there is no 'independent and already given criterion of what is just', while in the latter case there is (Rawls 1980: 523). But the contrast is overdrawn in this case, since, as we will see, there are some 'independent and already given criteria' for assessing the results of the original position. In any event, the two cases share the feature I am drawing attention to—the use of ignorance to ensure unbiased decisions.

4. Rawls denies that there is any essential similarity between his contractualism and Hare's impartial sympathizer. But as Barry puts it, this denial 'seems to me simply a flailing of the air' (Barry 1989: 410 n. 30). It is unfortunate that Rawls exaggerates the distance between his theory and Hare's, for the exaggeration works to Rawls's disadvantage. See the discussion of feminist critiques of Rawls in ch. 7, s. 3c(ii) below.

5. This objection is raised by Barry and Sen, although they mistakenly blame the problem on Rawls's commitment to using primary goods to define the least well-off position (Barry 1973: 55–7; Sen 1980: 215–16). The problem actually lies in Rawls's incomplete use of primary goods—i.e. his arbitrary exclusion of natural primary goods from the index. Rawls does discuss the idea of compensating natural disadvantages, but only in terms of a 'principle of redress' under which compensation is made in order to remove the direct effects of the handicap and thereby create equality of opportunity (Rawls 1971: 100–2). Rawls rightly rejects this view as both impossible and undesirable. But why not view compensation as a way of eliminating an undeserved inequality in overall primary goods? Compensating people for the unchosen costs of their natural disadvantages should be done, not so that they can compete with others on an equal footing, but so they can have the same ability to lead a satisfying life. For more on this, compare Michelman (1975: 330–9), Gutmann (1980: 126–7), and Daniels (1985: ch. 3) with Pogge (1989: 183–8) and Mapel (1989: 101–6).

Some commentators argue that Rawls does support compensating natural disadvantages, but not as a matter of justice. Instead he views our obligations to the naturally disadvantaged as 'duties of public benevolence' (Martin 1985: 189–91) or 'claims of morality' (Pogge 1989: 186–91, 275). These obligations to the disadvantaged are not matters of mere charity, for they should be compulsorily enforced through the state, but nor are they claims of justice. According to Pogge and Martin, Rawls's theory of justice is about 'fundamental justice', whereas compensation for the naturally disadvantaged is about 'the overall fairness of the universe' (Martin 1985: 180; Pogge 1989: 189). Unfortunately, neither author explains this contrast, nor how it is consistent with Rawls's emphasis on 'mitigating the effects of natural accident and social fortune' (Rawls 1971: 585). Martin, for example, seems to say that mitigating the effects of differential natural *assets* is a matter of fundamental justice, whereas mitigating the effects of differential natural *handicaps* is a matter of benevolence (Martin 1985: 178). It is hard to see what, within a Rawlsian approach, justifies this distinction. (Brian Barry argues that this restriction is only legitimate if Rawls is abandoning the whole idea of justice as equal consideration and adopting instead the Hobbesian idea of justice as mutual advantage—Barry 1989: 243–6; cf. n. 2 above.)

6. It is not impossible to imagine people who will object even when the envy test is met. Since the envy test says nothing about people's welfare, it is possible that, of two equally talented people, one will be miserable while the other is elated. All the envy test tells us is that the miserable person would be even more miserable if he had the resource bundle that the elated person possesses. Imagine someone who is congenitally moody and taciturn, regardless of the sort of resources he has and the sort of success he has in his projects. In this case, satisfying the envy test will not yield equal benefits to each person. Since the miserable person cannot control his congenital grumpiness, we might think that he therefore has some extra claim on resources. (On the other hand, since the person's misery is *ex hypothesi* not due to the bundle of resources he has, it is not clear how any redistribution will change his misery.)

This example suggests that the simple typology Dworkin works with is inadequate. He tends to view everything as either ambitions (which he sees as co-terminous with our personality-manifesting choices), or resources (which he sees as matters of unchosen circumstance). But there are personal attributes or psychological propensities (like grumpiness) which do not fit easily in either category, yet which affect how much benefit people get from social resources. For a critique of Dworkin's categories, see Cohen (1989: 916–34); Arneson (1989); Alexander and Schwarzschild (1987: 99); Roemer (1985*a*). While I cannot discuss these cases in depth, I believe that they (and other difficult cases such as uncontrollable cravings) complicate, rather than undermine, the aims and methods of Dworkin's theory. (As Dworkin notes, cravings or congenital moodiness can be viewed as a kind of natural disadvantage which could be insured against, along with other mental and physical disabilities—Dworkin 1981: 301–4.)

7. There may be a better middle ground between ignoring and equalizing circumstances than Dworkin's insurance scheme. Amartya Sen's 'equality of capacities' scheme is one possibility, which Rawls himself seems to endorse for the handicapped (Rawls 1982b: 168; cf. Sen 1980: 218–19). Sen aims at a kind of equalization for those with natural disadvantage, but he limits it to the equalization of 'basic capacities', rather than the full-fledged equalization of circumstances which Dworkin rejects as impossible. The extent to which this is possible, or different from the results of Dworkin's insurance scheme, is difficult to tell (Cohen 1989: 942; cf. Sen 1985: 143–4; 1990: 115 n. 12).

8. Whereas Dworkin argues that a just distribution would require more welfare redistribution than is currently provided, Rawls argues that a just distribution would involve less. He seems to think that market incomes in a property-owning democracy will naturally satisfy the difference principle (Rawls 1971: 87), and indeed will correspond to Dworkin's idea of an ambition-sensitive, endowment-insensitive distribution (Rawls 1971: 305; cf. DiQuattro 1983: 62–3). Hence he opposes progressive income tax, and the extensive redistribution of market income (Rawls 1971: 278–9). Like Mill, Rawls seems to think that welfare provision would be 'of very minor importance' were 'the diffusion of property satisfactory' (Mill 1965: 960). But if Dworkin neglects the need to distribute property equally, Rawls neglects the need to redistribute incomes fairly. For even in his property-owning democracy there will be undeserved differences in market income due to differential natural talents, and undeserved differences in needs due to natural disadvantages and other forms of misfortune (Krouse and McPherson 1988: 94–9; Carens 1985: 49–59; 1986: 40–1).

This points to another interesting difference between Rawls and Dworkin. Rawls thinks that the difference principle will, in practice, be similar to Dworkin's ambition-sensitive, endowment-insensitive distributive ideal, since the market naturally generates such a distribution. Dworkin thinks that the ambition-sensitive, endowment-insensitive ideal will, in practice, be similar to Rawls's difference principle, since neither markets nor governments can distinguish endowments and ambitions. Hence they both claim that their theory will, in practice, be similar to the other's, but for opposite reasons.

9. I have primarily been concerned to show that the liberal egalitarian view of an ideally just society endorses some fairly radical goals. It is a further question whether liberals should adopt radical means to achieve

such goals. On this question, Rawls and Dworkin are explicitly reformist rather than revolutionary. They both argue that respect for people's liberty takes precedence over, and puts limits on, the pursuit of a just distribution of material resources (Rawls 1971: 303; 1982*b*: 11; Dworkin 1987: 48–9). I cannot discuss this issue here, although these stipulations seem quite arbitrary, and unjustified in terms of the motivations of Rawls's contractors (Pogge 1989: 127–48).

4

Libertarianism

1. PROPERTY-RIGHTS AND THE FREE MARKET

(a) The diversity of right-wing political theory

Libertarians defend market freedoms, and demand limitations on the use of the state for social policy. Hence they oppose the use of redistributive taxation schemes to implement a liberal theory of equality. But not everyone who favours the free market is a libertarian, for they do not all share the libertarian view that the free market is inherently just. For example, one common argument for unrestricted capitalism is its productivity, its claim to be maximally efficient at increasing social wealth. Many utilitarians, convinced of the truth of that claim, favour the free-market, since its efficiency allows for the greatest overall satisfaction of preferences (Barry 1986: chs. 2–4). But the utilitarian commitment to capitalism is necessarily a contingent one. If, as most economists agree, there are circumstances where the free market is not maximally productive—e.g. cases of natural monopolies—then utilitarians would support government restrictions on property-rights. Moreover, some utilitarians argue that redistribution can increase overall utility even when it decreases productivity. Because of declining marginal utility, those at the bottom gain more from redistribution than those at the top lose, even when redistribution lessens productivity.

Others defend capitalism on the grounds not of maximizing utility, but of minimizing the danger of tyranny. Giving governments the power to regulate economic exchanges centralizes power, and since power corrupts, market regulations are the first step on 'the road to serfdom', in Hayek's memorable phrase. The more governments are able to control economic life, the more able (and willing) they will be to control all aspects of our lives. Hence capitalist freedoms are needed to preserve our civil and political liberties (Hayek 1960: 121; Gray 1986a: 62–8; 1986b: 180–5). But this defence of market freedom must also be a contingent one, for history does not reveal any invariable link between capitalism and civil liberties. Countries with essentially unrestricted capitalism have sometimes had poor human rights records (e.g. McCarthyism in the United States), while countries with an extensive welfare state have sometimes had excellent records in defending civil and political rights (e.g. Sweden).

So these two defences of the free market are contingent ones. More importantly, they are instrumental defences of the free market. They tell us that market freedoms are a means for promoting maximal utility, or for protecting political and civil liberties. On these accounts, we do not favour the free market because people have rights to property. Rather we give people property-rights as a way of increasing utility or stabilizing democracy, and if we could promote utility or stability some other way, then we could legitimately restrict property-rights.

Libertarianism differs from other right-wing theories in its claim that redistributive taxation is inherently wrong, a violation of people's rights.[1] People have a right to dispose of their goods and services freely, and they have this right whether or not it is the best way to ensure productivity. Put another way, government has no right to interfere in the market, even in order to increase efficiency. As Robert Nozick puts it, 'Individuals have rights, and there are things no person or group may do to them (without violating their rights). So strong and far-reaching are these rights that they raise the question of what, if anything, the state and its officials may do' (Nozick 1974: ix). Because people have a right to dispose of their holdings as they see fit, government interference is equivalent to forced labour—a violation, not of efficiency, but of our basic moral rights.

(b) Nozick's 'entitlement theory'

How do libertarians relate justice and the market? I will focus on Nozick's 'entitlement theory'. The central claim in Nozick's theory, as in most other libertarian theories, is this: if we assume that everyone is entitled to the goods they currently possess (their 'holdings'), then a just distribution is simply whatever distribution results from people's free exchanges. Any distribution that arises by free transfers from a just situation is itself just. For the government to tax these exchanges against anyone's will is unjust, even if the taxes are used to compensate for the extra costs of someone's undeserved natural handicaps. The only legitimate taxation is to raise revenues for maintaining the background institutions needed to protect the system of free exchange—e.g. the police and justice system needed to enforce people's free exchanges.

More precisely, there are three main principles of Nozick's 'entitlement theory':

1. a principle of transfer—whatever is justly acquired can be freely transferred;
2. a principle of just initial acquisition—an account of how people come initially to own the things which can be transferred in accordance with (1);
3. a principle of rectification of injustice—how to deal with holdings if they were unjustly acquired or transferred.

If I own a plot of land, then (1) says that I am free to engage in any transfers I wish to make concerning that land. Principle (2) tells us how the land initially came to be owned. Principle (3) tells us what to do in the event that (1) or (2) is violated. Taken together, they imply that if people's current holdings are justly acquired, then the formula for a just distribution is 'From each as they choose, to each as they are chosen' (Nozick 1974: 160).

The conclusion of Nozick's entitlement theory is that 'a minimal state, limited to the narrow functions of protection against force, theft, fraud, enforcement of contracts, and so on, is justified; any more extensive state will violate persons' rights not to be forced to do certain things, and is unjustified' (Nozick 1974: ix). Hence there is no public education, no public health care, transportation, roads, or parks. All of these involve the coercive taxation of some people against their will, violating the principle 'from each as they choose, to each as they are chosen'.

As we have seen, Rawls and Dworkin also emphasize that a just distribution must be sensitive to people's choices. But they believe that this is just half of the story. A just distribution must be ambition-sensitive, as Nozick's is, but it must also be endowment-insensitive, as Nozick's is not. It is unfair for the naturally disadvantaged to starve just because they have nothing to offer others in free exchange, or for children to go without health care or education just because they were born into a poor family. Hence liberal egalitarians favour taxing free exchanges in order to compensate the naturally and socially disadvantaged.

Nozick says this is unjust, because people are entitled to their holdings (if justly acquired), where 'entitled' means 'having an absolute right to freely dispose of it as one sees fit, so long as it does not involve force or fraud'. There are some limits on what I can do—my entitlement to my knife does not include the right to deposit it in your back, since you are entitled to your back. But otherwise I am free to do what I want with my resources—I can spend them on acquiring the goods and services of others, or I can simply give them away to others (even to the government), or I can decide to withhold them from others (including the government). No one has the right to take them from me, even if it is to keep the disabled from starving.

Why should we accept Nozick's claim that people's property entitlements are such as to preclude a liberal redistributive scheme? Some critics argue that Nozick has no argument—he gives us 'libertarianism without foundations' (Nagel 1981). But a more generous reading will detect two different arguments. As with Rawls, the first argument is an intuitive one, trying to draw out the attractive features of the free exercise of property-rights. The second is a more philosophical argument which attempts to derive property-rights from the premiss of 'self-ownership'. In line with my general approach, and I think with Nozick's intentions, I will interpret this self-ownership argument as an appeal to the idea of treating people as equals.

Other writers defend libertarianism by quite different arguments. Some libertarians argue that Nozick's entitlement theory is best defended by an appeal to liberty, rather than equality, while others attempt to defend it by an appeal to mutual advantage, as expressed in a contractarian theory of rational choice. So, in addition to Nozick's arguments, I will examine the idea of a right to liberty (s. 4), and the contractarian idea of mutual advantage (s. 3).

(c) The intuitive argument: the Wilt Chamberlain example

Firstly, then, Nozick's intuitive argument. As we have seen, his 'principle of transfer' says that if we have legitimately acquired something, we have absolute property-rights over it. We can freely dispose of it as we see fit, even though the effect of these transfers is likely to be a massively unequal distribution of income and opportunity. Given that people are born with different natural talents, some people will be amply rewarded, while those who lack marketable skills will get few rewards. Due to these undeserved differences in natural talents, some people will flourish while others starve. And these inequalities will then affect the opportunities of children, some of whom are born into privileged circumstances, while others are born into poverty. These inequalities, which Nozick concedes are possible results of unrestrained capitalism, are the source of our intuitive objections to libertarianism.

How then can Nozick hope to provide an intuitive defence of these rights? He asks us to specify an initial distribution which we feel is legitimate, and then argues that we intuitively prefer his principle of transfer to liberal principles of redistribution as an account of what people can legitimately do with their resources. Let me quote his argument at length:

> It is not clear how those holding alternative conceptions of distributive justice can reject the [entitlement theory]. For suppose a distribution favoured by one of these non-entitlement conceptions is realized. Let us suppose it is your favourite one and let us call this distribution D1; perhaps everyone has an equal share, perhaps shares vary in accordance with some dimension you treasure. Now suppose that Wilt Chamberlain is greatly in demand by basketball teams, being a great gate attraction. . . . He signs the following sort of contract with a team: In each home game, twenty-five cents from the price of each ticket of admission goes to him. . . . The season starts, and people cheerfully attend his team's games; they buy their tickets, each time dropping a separate twenty-five cents of their admission price into a special box with Chamberlain's name on it. They are excited about seeing him play; it is worth the total admission price to them. Let us suppose that in one season one million persons attend his home games, and Wilt Chamberlain winds up with $250,000, a much larger sum than the average income and larger even than anyone else has. Is he entitled to this income? Is this new distribution D2, unjust? If so, why? There is *no* question about whether each of the people was entitled to the control over the resources they held in D1; because that was the distribution (your favorite) that (for the purposes of argument) we assumed was acceptable. Each of these persons *chose* to give twenty-five cents of their money to Chamberlain. They could have spent it on going to the movies, or on candy bars, or on copies of *Dissent* magazine, or of *Monthly Review*. But they all, at least one million of them, converged on giving it to Wilt Chamberlain in exchange for watching him play basketball. If D1 was a just distribution, and people voluntarily moved from it to D2, transferring parts of their

shares they were given under D1 (what was it for if not to do something with?), is not D2 also just? If the people were entitled to dispose of the resources to which they were entitled (under D1), did not this include their being entitled to give it to, or exchange it with, Wilt Chamberlain? Can anyone else complain on grounds of justice? Each other person already has his legitimate share under D1. Under D1, there is nothing that anyone has that anyone else has a claim of justice against. After someone transfers something to Wilt Chamberlain, third parties *still* have their legitimate shares; *their* shares are not changed. By what process could such a transfer among two persons give rise to a legitimate claim of distributive justice on a portion of what was transferred, by a third party, who had no claim of justice on any holding of the others *before* the transfer? (1974: 160–2)

Because D2 seems legitimate, Nozick argues, his principle of transfer is more in line with our intuitions than redistributive principles like Rawls's difference principle.

What are we to make of this argument? It has some initial attraction because it emphasizes that the whole point of having a theory of fair shares is that it allows people to do certain things with them. It is perverse to say that it is very important that people get their fair share, but then prevent people from using that share in the way they desire. But does this confront our intuition about undeserved inequalities? Let us assume that I specified an initial distribution D1 that was in line with Rawls's difference principle. Hence each person starts with an equal share of resources, regardless of their natural talents. But at the end of the basketball season, Chamberlain will have earned $250,000, while the handicapped person, who may have no earning power, will have exhausted her resources, and will be on the verge of starvation. Surely our intuitions still tell us that we can tax Chamberlain's income to prevent that starvation. Nozick has persuasively drawn on our intuition about acting on our choices, but his example ignores our intuition about dealing fairly with unequal circumstances.

Indeed when Nozick does confront the question of unequal circumstances, he concedes the intuitive strength of the liberal position. He admits that it seems unfair for people to suffer undeserved inequalities in their access to the benefits of social co-operation. He 'feels the power' of this objection. However,

The major objection to speaking of everyone's having a right *to* various things such as equality of opportunity, life, and so on, and enforcing this right, is that these 'rights' require a substructure of things and materials and actions; and *other* people may have rights and entitlements over these. No one has a right to something whose realization requires certain uses of things and activities that other people have rights and entitlements over. (1974: 237–8)

In other words, we cannot tax Wilt Chamberlain to compensate for people's handicaps because he has absolute rights over his income. But Nozick concedes that our intuitions do not uniformly favour this account of property-rights. On the contrary, he accepts that some of our most powerful intuitions favour compensating undeserved inequalities. The problem with fulfilling that intuitively attractive idea, however, is that people have rights over their income. While Mackie's idea of a general right to 'a fair go' in life is intuitively attractive, 'the particular rights over things fill the space of rights, leaving no room for general rights to be in a certain material condition' (1974: 238).

Do 'particular rights over things fill the space of rights', leaving no room for a right to a fair go in life? I will examine Nozick's second, more philosophical, argument for that view later. But his attempt to provide an intuitive defence of it through the Wilt Chamberlain example is misleading. To see this more clearly, we can separate theories of distributive justice into three elements (cf. van der Veen and Van Parijs 1985: 73).

(P) Moral principles (e.g. Nozick's principle of 'self-ownership', or Rawls's principle of the 'moral arbitrariness' of natural talents);

(R) Rules of justice that govern the basic structure of society (e.g. Nozick's three rules of justice in appropriation, transfer, and rectification, or Rawls's 'difference principle');

(D) A particular distribution of holdings in a given time and place (e.g. which particular people are currently entitled to which particular resources).

The moral principles (P) define the rules of justice (R), which in turn generate a particular distribution (D).

What Nozick hopes to do in the Chamberlain example is support his account of moral principles (P) and just rules (R) by showing that we intuitively support a distribution (D2) which is generated by those rules. Even though the initial distribution D1 was generated by a different set of rules and principles (Rawls's difference principle, in my case), Nozick claims that we intuitively accept a subsequent distribution D2 that is generated by his rule of justice in transfer. But in fact Nozick's argument only seems to work because he interprets the initial distribution (D1) in terms of his own account of principles (P) and just rules (R). While Nozick allows us to specify the initial distribution of holdings, he assumes that we are thereby distributing full property-rights over these holdings, as required by his preferred theory of just rules. But this assumption is misleading, since our preferred theory of just rules may not involve distributing such particular rights to particular people.

For example, the reason I suggested a D1 based on Rawls's theory is that it removes undeserved disadvantages in people's circumstances. Giving particular people access to particular resources is a way of implementing the more general right to a fair go in life that underlies Rawls's theory. That very motivation for D1 would also give me a reason to put limits on the way that resources can be transferred. For example, I would put a redistributive taxation scheme in place as a way of continuing to mitigate the effects of undeserved natural disadvantages after that initial distribution. I would include that redistributive scheme alongside the initial distribution because my motivation in specifying D1 was not to give 'particular rights over particular things [to] particular people'. It was, rather, to implement some more general right to a fair go in life. D1 was my preferred distribution because it was generated by my preferred R (just rules), which in turn reflect my fundamental moral premisses (P) about moral equality, undeserved rewards, etc. And just as D1 was generated by my preferred conception of R and P, so I would want any distributions which come out of D1 to be consistent with them—i.e. to respect people's right to a fair go in life.

Nozick twists this around. He takes my D1 as specifying a set of absolute rights over particular things. He then says that because particular people have absolute rights over these particular things, therefore we cannot use redistributive taxation to meet the general right to a fair go. Whereas I gave particular people access to particular resources in order to implement a more general right to a fair go in life, Nozick makes it seem as if I gave particular people rights to particular things that prevent the implementation of a general right to a fair go. He thus distorts what I allowed in D1, and why I allowed it. I recommended that people have some control over resources in D1, because that distribution dealt fairly with unequal circumstances. Nozick says that I gave absolute control over the resources, and uses that fact to block attempts to deal fairly with unequal circumstances. It is of course true that if people were accorded absolute rights over the particular things distributed in D1, then it would be wrong to tax Chamberlain's income in order to support the disadvantaged. But I did not accord such rights, and the fact they would prevent us from keeping the disadvantaged from starving is a very good reason why I did not.

If we realize what Nozick is up to, then we can respond in a different way to his example. The best response to his offer to specify D1 is to refuse to specify any distribution at all. For if Nozick insists on treating D1 as endowing absolute rights, then we may not believe there is a fair initial distribution of such rights. If we realize that Nozick is saying, 'Here are some absolute rights to property—distribute them as you like', then we should politely refuse his offer. For the legitimacy of such rights is precisely what is in question.

2. THE SELF-OWNERSHIP ARGUMENT

(a) The principle of self-ownership

The Wilt Chamberlain example reveals the implausibility of defending libertarianism by a simple appeal to our intuitions about justice. A successful defence, therefore, will have to show that libertarianism, despite its unattractive features, is the unavoidable consequence of some deeper principle that we are strongly committed to. Libertarians differ, however, on what this deeper principle is. Some libertarians appeal to a principle of mutual advantage, others to a principle of liberty. I will examine these two defences later. Nozick, on the other hand, appeals to a principle of 'self-ownership', which he presents as an interpretation of the principle of treating people as 'ends in themselves'. This principle of treating people as ends in themselves, which was Kant's formula for expressing our moral equality, is also invoked by Rawls, and by utilitarians. It is indeed a principle to which we are strongly committed, and if Nozick can show that it yields self-ownership, and that self-ownership yields libertarianism, then he will have provided a strong defence of libertarianism. I will argue, however, that Nozick fails to derive either self-ownership or property-ownership from the idea of treating people as equals, or as ends in themselves.[2]

The heart of Nozick's theory, laid out in the first sentence of his book, is that 'Individuals have rights, and there are things no person or group may do to them (without violating their rights)' (1974: ix). Society must respect these rights because they 'reflect the underlying Kantian principle that individuals are ends and not merely means; they may not be sacrificed or used for the achieving of other ends without their consent' (1974: 30–1). This 'Kantian principle' requires a strong theory of rights, for rights affirm our 'separate existences', and so take seriously 'the existence of distinct individuals who are not resources for others' (1974: 33). Because we are distinct individuals each with our own distinct claims, there are limits to the sacrifices that can be asked of one person for the benefit of others, limits that are expressed by a theory of rights. This is why utilitarianism, which denies the existence of such limits, is unacceptable to Nozick. Respecting these rights is a necessary aspect of respecting people's claim to be treated as ends in themselves, not means for others. According to Nozick, a libertarian society treats individuals, not as 'instruments or resources', but as 'persons having individual rights with the dignity this constitutes. Treating us with respect by respecting our rights, it allows us, individually or with whom we choose, to choose our life and to realize our ends and our conception of ourselves, in so far as we can, aided by the voluntary co-operation of other individuals possessing the same dignity' (1974: 334).

There are important continuities here between Nozick and Rawls, not only in Nozick's appeal to the abstract principle of equality, but also in his more specific arguments against utilitarianism. It was an important part of Rawls's argument that utilitarianism fails to treat people as ends in themselves, since it allows some people to be sacrificed endlessly for the greater benefit of others. So both Rawls and Nozick agree that treating people as equals requires limits on the ways that one person can be used for the benefit of others, or for the benefit of society generally. Individuals have rights which a just society will respect, rights which are not subject to, or the product of, utilitarian calculations.

Rawls and Nozick differ, however, on the question of which rights are most important in treating people as ends in themselves. To over-simplify, we can say that for Rawls, one of the most important rights is a right to a certain share of society's resources. For Nozick, on the other hand, the most important rights are rights over oneself—the rights which constitute 'self-ownership'. The idea of having ownership rights over oneself may seem bizarre, as it suggests that there is a distinct thing, the self, which one owns. But the term 'self' in self-ownership has a 'purely reflexive significance. It signifies that what owns and what is owned are one and the same, namely, the whole person' (Cohen 1986a: 110). The basic idea of self-ownership can be understood by comparison with slavery—to have self-ownership is to have the rights over one's person that a slaveholder has over a chattel slave.

It is not immediately apparent what this difference amounts to. Why cannot we accept both positions? After all, the claim that we own ourselves does not yet say anything about owning external resources. And the claim that we have rights to a fair share of society's resources does not seem to preclude the possibility that we own ourselves. Nozick believes, however, that the two are not compatible. According to Nozick, Rawls's demand that goods produced by the talented be used to improve the well-being of the disadvantaged is incompatible with recognizing self-ownership. If I own my self, then I own my talents. And if I own my talents, then I own whatever I produce with my self-owned talents. Just as owning a piece of land means that I own what is produced by the land, so owning my talents means that I own what is produced by my talents. Hence the demand for redistributive taxation from the talented to the disadvantaged violates self-ownership.

The problem is not that Rawls and Dworkin believe that other people can own me or my talents, the way that a slave is owned by another person. On the contrary, as I have tried to show, their hypothetical positions are intended to model the claim that no one is the possession of any other (ch. 3, s. 3 above). There are many ways in which liberals respect individuals' claims over their own talents. Liberals accept that I am the legitimate possessor of my talents, and that I am free to use them in accordance with my chosen projects. However, liberals say that because it is a matter of brute luck that people have the talents they do, their rights over their talents do not include the right to accrue unequal rewards from the exercise of those talents. Because talents are undeserved, it is not a denial of moral equality for the government to consider people's talents as part of their circumstances, and hence as a possible ground for claims to compensation. People who are born naturally disadvantaged have a legitimate claim on those with advantages, and the naturally advantaged have a moral obligation to the disadvantaged. Thus, in Dworkin's theory, the talented owe insurance premiums that get paid out to the disadvantaged, while in Rawls's theory, the talented only benefit from their talents if it also benefits the disadvantaged.

For Nozick, this constitutes a denial of self-ownership. I cannot be said to own my talents if others have a legitimate claim on the fruits of those talents. Rawls's principles 'institute (partial) ownership by others of people and their actions and labour. These principles involve a shift from the classical liberals' notion of self-ownership to a notion of (partial) property-rights in *other* people' (Nozick 1974: 172). According to Nozick, this liberal view fails to treat people as equals, as ends in themselves. Like utilitarianism, it makes some people mere resources for the lives of others, since it makes some part of them (i.e. their natural attributes) a resource for all. Since I have rights of self-ownership, the naturally disadvantaged have no legitimate claim over me or my talents. The same is true of all other coercive interventions in free-market exchanges. Only unrestricted capitalism can fully recognize my self-ownership.

We can summarize Nozick's argument in two claims:

1. Rawlsian redistribution (or other coercive government interventions in market exchanges) is incompatible with recognizing people as self-owners. Only unrestricted capitalism recognizes self-ownership.
2. recognizing people as self-owners is crucial to treating people as equals.

Nozick's conception of equality begins with rights over one's self, but he believes that these rights have implications for our rights to external resources, implications that conflict with liberal redistribution.

This is an untenable position, for two reasons. Firstly, Nozick is mistaken in believing that self-ownership necessarily yields absolute property-rights. Self-ownership is compatible with various regimes of property-ownership, including a Rawlsian one. Secondly, the principle of self-ownership is an inadequate account of treating people as equals, even on Nozick's own view of what is important in our lives. If we try to reinterpret the idea of self-ownership to make it a more adequate conception of equality, and select an economic regime on that basis, we will be led towards, rather than away from, the liberal view of justice.

(b) Self-ownership and property-ownership

I will examine these two objections in turn. Firstly, how does self-ownership yield property-ownership? Nozick claims that market exchanges involve the exercise of individuals' powers, and since individuals own their powers, they also own whatever comes from the exercise of those powers in the marketplace.

But this is too quick. Market exchanges involve more than the exercise of self-owned powers. They also involve legal rights over things, over external goods, and these things are not just created out of nothing by our self-owned powers. If I own some land, I may have improved the land, through the use of my self-owned powers. But I did not create the land, and so my title to the land (and hence my right to use the land in market exchanges) cannot be grounded solely in the exercise of my self-owned powers.

Nozick recognizes that market transactions involve more than the exercise of self-owned powers. In his theory, my title to external goods like land comes from the fact that others have transferred the title to me, in accordance with the principle of transfer. This assumes, of course, that the earlier owner had legitimate title. If someone sells me some land, my title to the land is only as good as her title, and her title was only as good as the one before her, and so on. But if the validity of my property-rights depends on the validity of previous property-rights, then determining the validity of my title over external goods requires going back down the chain of transfers to the beginning. But what is the beginning? Is it the point where someone created the land with their self-owned powers? No, for no one created the land. It existed before human beings existed.

The beginning of the series of transfers is not when the land was created, but rather when it was first appropriated by an individual as her private property. On Nozick's theory, we must go down the chain of transfers to see if the initial acquisition was legitimate. And nothing in the fact, if it is a fact, that we own our talents ensures that anyone can legitimately appropriate for themselves something they did not create with their talents. If the first person who took it did so illegitimately, then she has no legitimate title over it, and hence no legitimate right to transfer it to someone else, who would then have no legitimate right to transfer it to me. Hence if I am to be entitled to all of the rewards which accrue to me from market exchanges, as Nozick believes I am, I must be the legitimate owner not only of my powers, but also of initially unowned external resources.

This question about the initial acquisition of external resources is prior to any question about legitimate transfer. If there was no legitimate initial acquisition, then there can be no legitimate transfer, on Nozick's theory. So Nozick owes us an account of how external resources come to be initially acquired by one individual for her own use. Nozick is aware that he needs such an account. There are times when he says that 'things come into the world already attached to people having entitlements over them' (1974: 160). But he realizes that everything which is owned today includes an element which did not come into the world as private property, legally or morally. Everything that is now owned has some element of nature in it. How then did these natural resources, which were not initially owned by anyone, come to be part of someone's private property?

(i) Initial acquisition

The historical answer is often that natural resources came to be someone's property by force, which raises a dilemma for those who hope Nozick's theory will defend existing inequalities. Either the use of force made the initial acquisition illegitimate, in which case current title is illegitimate, and there is no moral reason why governments should not confiscate the wealth and redistribute it. Or the initial use of force did not render the acquisition illegitimate, in which case we can, with equal justification, use force to take it away from its current owners and redistribute it. Either way, the fact that initial acquisition often involved force means that there is no moral objection to redistributing existing wealth (Cohen 1988: 253–4).

Nozick's response to this problem is the first one. The use of force makes acquisition illegitimate, so current title is illegitimate (1974: 230–1). Hence those who currently possess scarce resources have no right to deprive others of access to them—e.g. capitalists are not entitled to deprive workers of access to the products or profits of the existing means of production. Ideally, the effects of the illegitimate acquisition should be rectified, and the resources restored to their rightful owner. However, it is often impossible to know who the rightful owners are—we do not know from whom the resources were illegitimately taken. Nozick suggests that we could rectify the illegitimacy of existing title by a one-time general redistribution of resources in accordance with Rawls's difference principle. Only after this redistribution will the libertarian principle of transfer hold. Where we do know the rightful owners, however, we should restore the resources to them. For example, Nozick's view supports returning much of New England to the American Indians, whose initial title was unjustly taken away (Lyons 1981).

This rejection of the legitimacy of current title is not a curiosity of Nozick's presentation that can be detached from the rest of his theory. If one really believes in Nozick's entitlement theory, then current title is only as legitimate as previous titles. If previous title was legitimate, then any new distribution which results from market exchanges is just. That is what libertarians propose as their theory of justice. But the corollary of that theory is that if previous title was illegitimate, so is the new distribution. The fact that the new distribution arose from market transactions is irrelevant, since no one had any right to transfer those resources through market exchanges. This, as much as the first case, is an essential part of Nozick's theory. They are two sides of the same coin.

Because most initial acquisition was in fact illegitimate, Nozick's theory cannot protect existing holdings from redistribution. But we still need to know how acquisition could have arisen legitimately. If we cannot answer that question, then we should not only postpone the implementation of Nozick's principle of transfer until historical titles are ascertained or rectified, we should reject it entirely. If there is no way that people can appropriate unowned resources for themselves without denying other people's claim to equal consideration, then Nozick's right of transfer never gets off the ground.

What sort of initial acquisition of absolute rights over unowned resources is consistent with the idea of treating people as equals?

This is an old problem for libertarians. Nozick draws on John Locke's answer to it. In seventeenth-century England there was a movement towards the 'enclosure' (private appropriation) of land which had previously been held in common for general use. This land ('the commons') had been available to all for the grazing of animals, or for gathering wood, etc. As a result of this private appropriation, some people became wealthy while others lost their access to resources, and so lost their ability to sustain themselves. Locke wished to defend this process, and so needed to give an account of how people come, in a morally legitimate way, to have full ownership rights over the initially unowned world.

Locke's answer, or at any rate one of his answers, was that we are entitled to appropriate bits of the external world if we leave 'enough and as good' for others. An act of appropriation that meets this criterion is consistent with the equality of other individuals since they are not disadvantaged by that appropriation. Locke also gave other answers—e.g. that we can appropriate that with which we have mixed our labour, so long as we do not waste it. But the 'enough and as good' criterion seems to do most of the work, even in Locke's own examples. He gives the example of picking up acorns where there are more than enough for everyone, or drinking water from a stream. In neither of these cases is there any real labour-mixture, and so long as enough and as good is left for others, who can object even if I waste some of the acorns or water? If my appropriation leaves everyone as well off as before, then who has been treated unjustly?

Locke realized that most acts of appropriation (unlike the two examples above) do not leave enough and as good of the object being appropriated. Those who enclosed the land in seventeenth-century England clearly did not leave enough and as good land for others. But Locke says that appropriation is acceptable if it leaves people as well or better off overall. While I have less land available to me, the result of enclosing the commons may be that many of the goods I buy will become cheaper, leaving me better off overall. So the test of a legitimate appropriation is that it does not worsen anyone's overall condition. Nozick calls this the 'Lockean proviso', and he adopts it as his test of legitimate acquisition: 'A process normally giving rise to a permanent bequeathable property right in a previously unowned thing will not do so if the position of others no longer at liberty to use the thing is thereby worsened' (1974: 178).[3] Emphasizing the question of whether the condition of others is worsened is appropriate, because it enforces the principle of equal consideration of people's interests. Acquisition does not violate equal consideration if it does not worsen anyone's situation.

What kinds of appropriation meet this test? It depends on what counts as being made worse off from an act of appropriation. Nozick's answer is that appropriation of a particular object is legitimate if its withdrawal from general use does not make people worse off in material terms than they had been when it was in general use. For example, consider Amy and Ben, who both live off land which is initially under general use. Amy now appropriates so much of the land that Ben cannot live off the remaining land. That might seem to make Ben worse off. But Amy offers Ben a wage to work on her land which exceeds what he was originally producing on his own. Amy also gets more resources than she initially produced, due to the increased productivity arising from a division of labour, and the increase in her share is larger than the increase in his share. Ben must accept this, since there is not enough land left for him to live as he used to. He needs access to the land that she appropriated, and she is able to dictate the terms of that access, so that he gets less than half of the benefits of the division of labour. Amy's act of appropriation satisfies Nozick's proviso, since the situation after her appropriation is better than general use in terms of material resources, for both Amy and Ben. (Actually, it need not be better for Ben, so long as it is not worse.)

In this way, the unowned world comes to be appropriated, with full property-rights, by self-owning people. Nozick believes that the proviso is easily met, and so most of the world will quickly come to be privately appropriated. Hence, self-ownership yields absolute ownership of the external world. Since initial appropriation includes the right of transfer, we will soon have a fully developed market for productive resources (i.e. the land). And since this appropriation excludes some people from access to those productive resources, we will soon have a fully developed market in labour. And since people will then legitimately own both the powers and the property which are involved in market exchange, they will be legitimately entitled to all the rewards which accrue from those exchanges. And since people will be entitled to all their market rewards, government redistribution to help the disadvantaged would be a violation of people's rights. It would be using some people as a resource for others.

(ii) The Lockean proviso

Has Nozick given us an acceptable account of fair initial acquisition? We can summarize it this way:

1. People own themselves.
2. The world is initially unowned.
3. You can acquire absolute rights over a disproportionate share of the world, if you do not worsen the condition of others.
4. It is relatively easy to acquire absolute rights over a disproportionate share of the world. Therefore:
5. once people have appropriated private property, a free market in capital and labour is morally required.

I will concentrate on Nozick's interpretation of (3), his account of what it is to worsen the condition of others. His account has two relevant features: (*a*) it defines 'worse off' in terms of material welfare; (*b*) it defines pre-appropriation common usage as the standard of comparison. I will argue that both of these features are inadequate, and the flaws are so serious that they require not just a modification of Nozick's test, but an abandonment of it. Any plausible test of initial acquisition will yield only limited property-rights.

Material welfare   The reason why Nozick emphasizes self-ownership, we have seen, is that we are separate individuals, each with our own life to live (s. 2*a* above). Self-ownership protects our ability to pursue our own goals, our 'conceptions of ourselves', since it allows us to resist attempts by others to use us merely as means to their ends. One would expect Nozick's account of what it is for an act of appropriation to worsen the condition of others similarly to emphasize people's ability to act on their conception of themselves, and to object to any appropriation that puts someone in an unnecessary and undesirable position of subordination and dependence on the will of others.

But notice that the fact that Ben is now subject to Amy's decisions is not considered by Nozick in assessing the fairness of the appropriation. In fact, Amy's appropriation deprives Ben of two important freedoms: (1) he has no say over the status of the land he had been utilizing—Amy unilaterally appropriates it without asking or receiving Ben's consent; (2) Ben has no say over how his labour will be expended. He must accept Amy's conditions of employment, since he will die otherwise, and so he must relinquish control over how he spends much of his time. Before the appropriation, he may have had a conception of himself as a shepherd living in harmony with nature. Now he must abandon those pursuits, and instead obey Amy's commands, which might involve activities that exploit nature. Given these effects, Ben may be made worse off by Amy's appropriating the land, even though it leads to a small increase in his material income.

Nozick should consider these effects, on his own account of why self-ownership is important. He says that the freedom to lead our lives in accordance with our own conception of the good is the ultimate value, so important that it cannot be sacrificed for other social ideals (e.g. equality of opportunity). He claims that a concern for people's freedom to lead their own lives underlies his theory of unrestricted property-rights. But his justification of the initial appropriation of property treats Ben's autonomy as irrelevant.

It is interesting that although Nozick claims that Ben is not made worse off by the appropriation, he does not require that Ben consent to the appropriation. If consent was required, Ben might well refuse. If Ben is right to refuse, since it really would make him worse off, then the appropriation should not be allowed. Perhaps Ben would be mistaken in refusing, since the gain in material welfare outweighs the loss of autonomy. In that case, we might allow Amy's appropriation as an act of paternalism. But Nozick claims to be against such paternalism. For example, he is against mandatory health insurance or pension plans that are instituted for people's own benefit. But appropriating private property can contradict a person's will as much as levying a tax on him can. It seems that Nozick opposes paternalism when it threatens property-rights, but willingly invokes it when it is required to generate property-rights. If we exclude paternalism, and emphasize autonomy, as Nozick himself does elsewhere in his theory, then justifying private appropriation becomes much more difficult (cf. Kernohan 1988: 70; Cohen 1986a: 127, 135).

Arbitrary narrowing of the options Nozick's proviso says that an act of appropriation must not make others worse off than they were when the land was in common use. But this ignores many relevant alternatives. Let us say that Ben, worried about the possibility of Amy unilaterally appropriating the land, decides to appropriate it for himself, and then offers Amy a wage to work on what is now his land, keeping to himself the bulk of the benefits of the increased productivity. This too passes Nozick's test. Nozick considers it irrelevant who does the appropriating, and who gets the profits, so long as the non-appropriator is not worsened by it. Nozick is, in effect, accepting a first-come, first-served doctrine of appropriation. But why should we accept this as a fair procedure for appropriation, rather than, for example, a system which equalizes chances for appropriation? Which is more in line with our intuitions about fairness, or with Nozick's own account of our interests? Should the most important value—our ability to lead our own lives—be dependent on the arbitrariness of a first-come, first-served doctrine?

Consider another alternative. This time Ben, who is a better organizer of labour, appropriates the land, and organizes an even greater increase in productivity, allowing both to get more than they got when Amy appropriated the land. They are both worse off when Amy appropriates than they would be when Ben appropriates. Yet Nozick allows Amy to appropriate, and denies that Ben is thereby made worse off, since he does better under Amy's appropriation than he did under common use of the land, which is the only alternative Nozick considers relevant.

Finally, what if Amy and Ben appropriate the land collectively, exercising ownership rights jointly, and dividing the labour consensually? If appropriation is going to take place amongst a community of self-owners, then Ben should have the option of collective ownership, rather than having Amy unilaterally deprive him of his ability to pursue his own conception of himself. Surely this is a relevant possibility, the existence of which puts in question the legitimacy of any act of unilateral appropriation that leaves others with insufficient access to resources to support themselves.

According to Nozick's proviso, all these alternatives are irrelevant. It does not matter to the legitimacy of an appropriation that some other appropriation better serves people's material interests, or their autonomy. But since all of these are genuine options, and since each option spares somebody a harm that will occur under Nozick's scheme, Nozick needs to explain why people are not made worse off when they are excluded. Unfortunately, Nozick simply ignores these possibilities.

These problems with Nozick's proviso are made clearer if we move to the level of capitalism as an ongoing system. The acts of initial appropriation which Nozick allows will quickly lead to a situation in which there are no more accessible useful unowned things. Those who were able to appropriate may have vast wealth, while others are entirely without property. These differences will be passed on to the next generation, some of whom will be forced to work at an early age, while others have all the privileges in life. This is acceptable to Nozick, so long as the system of appropriation and transfer continues to meet the Lockean proviso—i.e. just as individual acts of initial appropriation are legitimate if they do not make people worse off than they were when the world was unowned, so capitalism as an ongoing system is just if no one is worse off than they would have been without privatization of the external world.

Nozick invokes familiar accounts of capitalism's productivity and wealth-creation to support the claim that capitalism passes this test (1974: 177). Notice, however, that capitalism passes that test even though the propertyless are dependent for their survival on those with property wanting to buy their labour, and even though some people may starve because no one does want to buy their labour. This is acceptable to Nozick since people who lack marketable skills would have starved anyway had the land remained unowned. The propertyless lack a just grievance because 'those propertyless persons who do manage to sell their labor power . . . will get at least as much and probably more in exchange for it than they could have hoped to get by applying it in a Lockean state of nature; and those propertyless persons whose labor power is not worth buying, although they might therefore, in Nozick's non-welfare state, die, . . . would have died in the state of nature anyway' (Cohen 1986*b*: 85 n. 11).

This is an absurdly weak requirement. It is not enough that unrestricted capitalism does not make people worse off than they would have been without any private appropriation. These are not the only two options that are relevant to judgements about the legitimacy of appropriation. It is absurd to say that a person who starves to death is not made worse off by Nozick's system of appropriation when there are other systems in which that person would not have died. Nozick's refusal to consider these other possibilities is arbitrary and unjust.

The test of a legitimate appropriation, if it is to treat each person with equal consideration, must consider all the relevant alternatives, keeping in mind people's interest in both material goods and autonomy. Can we modify the Lockean proviso to include these considerations, while retaining its intuitive point that the test of appropriation is whether it worsens someone's condition? We might say that a system of appropriation worsens someone's condition if there is another possible scheme in which they would do better. Unfortunately, every system of property allocation will fail that test. The person who lacks marketable skills would be worse off in Nozick's pure capitalism than he would be under Rawls's difference principle; the person with marketable talents would be worse off under Rawls's regime than under Nozick's. In any given system, there will be someone who would do better in another system. That test is unreasonable anyway, for no one has a legitimate claim that the world be maximally adapted to suit their preferences. The fact that there is a possible arrangement in which I would be better off does not show that the existing system has harmed me in any morally significant sense. We want to know whether a system of appropriation makes people worse off, not compared to a world which is maximally adapted to their particular interests, but compared to a world in which their interests are fairly attended to.

It is an insufficient test of justice that people benefit relative to the initial state of common holdings. But nor can people demand that they have whatever system benefits them the most. The proviso requires a middle ground. It is difficult to say what that middle ground is, or how different it would be from the principles of Rawls and Dworkin. John Arthur argues that the appropriate test is an egalitarian one—appropriation worsens someone's condition if, as a result, they get less than an equal share of the value of the world's natural resources. This is the only decision that makes sense, he argues, 'in light of the fact that [each person] is as entitled to the resources as anybody else. He wasn't born deserving a smaller share of the earth's wealth, nor is anybody else naturally entitled to a larger than average share' (Arthur 1987: 344; cf. Steiner 1977: 49).

Cohen argues that Rawls's difference principle might provide a fair standard of legitimate appropriation (Cohen 1986*a*: 133–4). Other possible tests would lead to different results, but no plausible test would allow Nozick's unrestricted property-rights.

If the proviso recognizes the full range of interests and alternatives that self-owners have, then it will not generate unrestricted rights over unequal amounts of resources. Some people will be made worse off in important ways, compared to morally relevant alternatives, by a system which allows people to appropriate unequal amounts of the external world. And if, as Nozick himself says, 'Each owner's title to his holding includes the historical shadow of the Lockean proviso on appropriation', then, on any plausible interpretation, 'the shadow thrown by [the proviso] so entirely envelops such titles as to render them indiscernible' (Steiner 1977: 48; Nozick 1974: 180). Any title that self-owning people have over unequal resources will be heavily qualified by the claims of the propertyless.

Initial ownership of the world There is another problem with Nozick's proviso that blocks the move from self-ownership to unrestricted capitalism. Recall my summary of Nozick's argument:

1. People own themselves.
2. The world is initially unowned.
3. You can acquire absolute rights over a disproportionate share of the world, if you do not worsen the condition of others.
4. It is relatively easy to acquire absolute rights over a disproportionate share of the world. Therefore:
5. once private property has been appropriated, a free market in capital and labour is morally required.

My first argument concerned Nozick's interpretation of (3), which turned out to be too weak, so that (4) is false. But there is a second problem. Why accept (2), the claim that the world is initially unowned, and hence up for grabs? Why not suppose that the world is jointly owned, such that each person has an equal veto over the disposal of the land (Exdell 1977: 146–9; Cohen 1986*b*: 80–7)? Nozick never considers this option, but others, including some libertarians, claim that it is the most defensible account of world-ownership (Locke himself believed that the world initially belonged to everyone, not no one, for God 'hath given to men the world in common'—cf. Christman 1986: 159–64).

What would happen if the world was jointly owned, and hence not subject to unilateral privatization? There are a variety of possible outcomes, but in general they will negate the inegalitarian implications of self-ownership. For example, the disadvantaged may be able to use their veto to bargain for a distributive scheme like Rawls's difference principle. We might end up in this way with a Rawlsian distribution, not because we deny self-ownership (such that the disadvantaged could have a direct claim on the advantaged), but because we are joint owners of the external world (such that the untalented can veto uses of the land that benefit the talented without also benefiting them). A similar result might also arise if we view the external world as neither up for grabs, nor jointly owned, but as divided equally amongst all the members of the human community (Cohen 1986*b*: 87–90).

All of these accounts of the moral status of the external world are compatible with the principle of self-ownership, since self-ownership says nothing about what kind of ownership we have over external resources. And indeed various libertarians have endorsed these other options.[4] Each of these options has to be evaluated in terms of the underlying values that Nozick professes to care about. Nozick does not begin this evaluation, but it is clear that absolute property-rights over unequal bits of the world are only secured if we invoke weak and arbitrary premises about appropriation and the status of the external world.

(c) Self-ownership and equality

I have tried to show that the principle of self-ownership does not by itself generate a moral defence of capitalism, since a capitalist requires not only ownership of her self, but also ownership of resources.[5] Nozick believes that self-ownership inevitably leads to unrestricted property-rights, but we are in fact confronted by a variety of economic regimes that are compatible with self-ownership, depending on our theory of legitimate appropriation, and our assumptions about the status of the external world. Nozick believes that self-ownership requires that people be entitled to all the rewards of their market exchanges, but different regimes vary in the extent to which they allow self-owning individuals to retain their market rewards. Some will allow the naturally talented to translate their natural advantages into unequal ownership of the external world (although not necessarily to the extent allowed by Nozick); others will redistribute market income so as to ensure that the naturally disadvantaged have equal access to resources (as in Rawls or Dworkin). Self-ownership is compatible with all these options.

Which would Nozick favour? We can assume that he would prefer those regimes which leave property-rights as unrestricted as possible. Can he give us any reason to prefer such libertarian regimes over liberal egalitarian ones? I can think of three arguments he might give. These arguments draw on aspects of, but also go beyond, the idea of self-ownership, since that idea by itself is insufficient to identify a just distribution. One argument concerns consent, the second concerns the idea of self-determination, the third concerns dignity.

Nozick might say that the choice of economic regime should be decided, if possible, by the consent of self-owning people. And, he might claim, self-owning people would all choose a libertarian regime, were it up to them. But that is wrong. As we have seen, Nozick's own scheme of acquisition depended on Ben not having to give his consent to Amy's appropriation. Different people would do best in different economic regimes, and so would consent to different regimes. One could try to ensure unanimous consent by seeking agreement behind a veil of ignorance, as Rawls does. But that will not help Nozick, since, as we have seen, that leads to liberal, not libertarian, results.

Secondly, Nozick might claim that the assumptions which lead to liberal results, while formally compatible with self-ownership, in fact undermine the value of self-ownership. For example, the assumption that the world is jointly owned, or that it should be collectively appropriated, would nullify the value of self-ownership. For how can I be said to own myself if I may do nothing without the permission of others? In a world of joint ownership, do not Amy and Ben jointly own not only the world but also in effect each other? Amy and Ben may have legal rights over themselves (unlike the chattel slave), but they lack independent access to resources. Their legal rights of self-ownership are therefore purely formal, since they require each other's permission whenever they wish to use resources in the pursuit of their goals. We should select a regime that contains not only formal self-ownership, but also a more substantive self-ownership that provides one with effective control over one's life.

Following Charles Fried, I will use the term 'self-determination' to describe this more substantive conception of self-ownership. He says that it requires a 'determinate domain . . . free of the claims of others' (Fried 1983: 55). Similarly, Jon Elster says that substantive self-ownership involves 'the right to choose which of one's abilities to develop' (Elster 1986: 101). Common to both these interpretations of substantive self-ownership is the idea that in the central areas of our life, in our most important projects, we should be free to act on our own conceptions of the good. Both argue that respecting self-determination is an important part of treating people as ends not means, as distinct individuals each with their own life to lead.

I think that Nozick appeals to both the formal and substantive conceptions of self-ownership. He explicitly defends the formal conception, dealing with legal rights over one's physical being. But at least part of Nozick's defence of formal self-ownership is that it promotes substantive self-ownership—it promotes our ability to act on our conception of ourselves. So it seems plausible that Nozick would endorse selecting the regime which best promotes substantive self-ownership (within the constraints imposed by formal self-ownership). While different economic regimes are compatible with formal self-ownership, he might argue that liberal regimes render self-ownership purely formal, whereas the more libertarian regimes ensure substantive self-ownership, since libertarian property-rights leave people free to act without others' permission.

But this will not work, for in a libertarian regime only some people can translate their formal self-ownership into substantive self-determination. Libertarians cannot guarantee each person substantive control over her life, and indeed, Nozick explicitly says that formal self-ownership is all that people can legitimately claim. He says that the worker who lacks any property, and who must sell her labour on adverse terms to the capitalist, has full self-ownership (1974: 262–4). She has full self-ownership even though, Nozick concedes, she may be forced to agree to whatever terms the capitalist is offering her in order to survive. The resulting 'agreement' might well, as in Victorian England, be essentially equivalent to the enslavement of the worker. The fact that the worker has formal rights of self-ownership means that she cannot be the legal possession of another person (unlike the chattel slave), but economic necessity may force her to agree to terms which are just as adverse.

Lack of property can be just as oppressive as lack of legal rights. As Mill put it:

No longer enslaved or made dependent by force of law, the great majority are so by force of poverty; they are still chained to a place, to an occupation, and to conformity with the will of an employer, and debarred by the accident of birth both from the enjoyments, and from the mental and moral advantages, which others inherit without exertion and independently of desert. That this is an evil equal to almost any of those against which mankind have hitherto struggled, the poor are not wrong in believing. (Mill 1967: 710)

The full self-ownership of the propertyless worker is no more substantive than the self-ownership of Amy or Ben in a world of collective ownership. Amy has no access to productive resources without Ben's permission, but the same is true of the worker who is dependent on the agreement of the capitalist. In fact, people in a situation of collective ownership have more real control over their lives, since Amy and Ben must strike a deal in order to use their resources, whereas a capitalist need not strike an agreement with any particular worker in order to survive, especially if the worker does not possess a talent required by the capitalist.

Libertarianism not only restricts the self-determination of the propertyless worker, it makes her a resource for others. Those who enter the market after others have appropriated all the available property are 'limited to gifts and jobs others are willing to bestow on them', and so 'if they are compelled to co-operate in the scheme of holdings, they are forced to benefit others. This forced compliance with the property system constitutes a form of exploitation and is inconsistent with the most basic of [Nozick's] root ideas, rendering as it does the latecomers mere resources for others' (Bogart 1985: 833–4).

What regime best promotes substantive self-ownership? Self-determination requires resources as well as rights over one's physical being. We are only able to pursue our most important projects, free from the demands of others, if we are not forced by economic necessity to accept whatever conditions others impose on us in return for access to needed resources. Since meaningful self-determination requires both resources and liberties, and since each of us has a separate existence, each person should have an equal claim to these resources and liberties.

But, if so, then the concern for self-determination leads us towards liberal regimes, not libertarian ones. Libertarians claim that liberal welfare programmes, by limiting property-rights, unduly limit people's self-determination. Hence the removal of welfare redistribution programmes (Nozick), or their limitation to an absolute minimum (Fried), would be an improvement in terms of self-determination. But that is a weak objection. Redistributive programmes do restrict the self-determination of the well off to a limited degree. But they also give real control over their lives to people who previously lacked it. Liberal redistribution does not sacrifice self-determination for some other goal. Rather, it aims at a fairer distribution of the means required for self-determination. Libertarianism, by contrast, allows undeserved inequalities in that distribution—its concern with self-determination does not extend to a concern for ensuring the fair distribution of the conditions required for self-determination. In fact, it harms those who most need help in securing those conditions. If each person is to be treated as an end in herself, as Nozick says repeatedly, then I see no reason for preferring a libertarian regime to a liberal redistributive one.

A liberal regime which taxes the unequal rewards of undeserved talents does limit some people's self-determination. But it is an acceptable limit. Being free to choose one's own career is crucial to self-determination, but being free from taxation on the rewards which accrue from undeserved natural talents is not. Even if one's income is taxed in accordance with Rawlsian principles, one still has a fair share of resources and liberties with which to control the essential features of one's life. Taxing income from the exercise of natural talents does not unfairly disadvantage anyone in their substantive self-ownership, their ability to act according to their conception of themselves.

Finally, Nozick might argue that welfare redistribution denies people's dignity, and this dignity is crucial to treating people as equals (e.g. Nozick 1974: 334). Indeed Nozick often writes as if the idea that other people have claims on the fruits of my talents is an assault on my dignity. But this is implausible. One problem is that Nozick often ties dignity to self-determination, so that it will be liberal regimes, not libertarian ones, which best promote each person's dignity. In any event, dignity is predicated on, or a by-product of, other moral beliefs. We only feel something to be an attack on our dignity if we are already convinced that it is wrong. Redistribution will feel like an assault on dignity only if we believe it is morally wrong. If we believe instead that redistribution is a required part of treating people as equals, then it will serve to promote, rather than attack, people's sense of equal dignity.

Libertarianism cannot be defended in terms of self-ownership, consent, self-determination, or dignity. All of these are either indeterminate or support liberal egalitarianism. Perhaps there is some other reason Nozick might give for endorsing libertarianism. It is hard to say, since he falsely believes that self-ownership requires libertarianism, and so did not consider the alternatives. But as it stands, Nozick fails to defend absolute property-rights, or the free-market system that is designed to respect such rights. Self-ownership does not preclude redistributive taxation, for many different economic regimes are formally compatible with self-ownership. And if we look beyond formal self-ownership to those regimes which best ensure substantive self-ownership, then Nozick has not given us any reason to prefer libertarian inequalities to liberal equality.

But why should we be concerned with formal self-ownership at all? In the above argument, I used the idea of substantive self-ownership as a test for deciding between those regimes that are compatible with formal self-ownership. But if we contrast these two conceptions, surely substantive self-determination is more fundamental. We do not endorse self-determination because it promotes formal self-ownership. Rather, we will endorse formal self-ownership in so far as it promotes self-determination. Indeed, as I mentioned earlier, Nozick himself sometimes treats the substantive conception as the more fundamental. So why not just start with self-determination as our preferred conception of treating people as equals? Rather than ask which of the regimes that are compatible with formal self-ownership best promotes self-determination, why not just ask which regime best promotes self-determination? It may be that the best regime, assessed in terms of self-determination, not only goes beyond formal self-ownership, but also limits it. In that case, formal self-ownership should give way to the substantive self-determination that really matters to us (Cohen 1986*b*: 86).

This seems so obviously preferable that an explanation is needed for Nozick's emphasis on formal self-ownership. One explanation is simply that Nozick needed it in order to defend property-rights. But there is a more generous explanation. Nozick, like the classical liberals, wants to articulate a conception of equality which denies that anyone is by nature or by right subordinate to another. No one is merely a resource for others, the way a slave is the resource of his owner. If slavery is the paradigm case of a denial of equality, it might seem that the best way to affirm equality is to give each person the legal rights over himself that slave-owners have over their slaves; the best way to prevent the enslavement of one person to another is to give each person ownership over himself. Unfortunately, the fact that I have legal rights of self-ownership does not mean that I have the ability to avoid what is in effect enslavement to another. Even if the capitalist does not have the same legal rights over me that slave-owners had over slaves, I may not have any real ability to decide on the nature and terms of my living. The best way to prevent the sort of denial of equality that occurs in slavery is not to reverse the legal rights involved, but rather to equalize the substantive control each person possesses, in the form of equal liberties and resources.

Nozick's emphasis on the idea of formal self-ownership may also be due to the undifferentiated nature of that concept. The idea of self-ownership misleadingly suggests that we either have or lack self-ownership, as if the various rights and powers which constitute self-ownership must be accepted or rejected as a package. If that was indeed our choice, then it would make sense to emphasize self-ownership. But in reality there is a range of options, involving different kinds of control over one's choices and one's circumstances. The idea of self-ownership tends to prevent people from considering all the relevant options, as Nozick's own discussion reveals. The claim that undifferentiated self-ownership is crucial to treating each person as an end in herself is only plausible if it is being compared with the single option of the undifferentiated denial of self-ownership.

We need to distinguish different elements involved in controlling one's self, and see how they relate to the different elements involved in controlling external resources. We should consider each of these rights and powers on its own terms, to see in what ways it promotes each person's essential interests. Which combination of rights and resources contributes to each person's ability to act on their goals and projects, their conception of themselves? The best mix will involve more than formal self-ownership (e.g. access to resources), but it may also involve less, for it may be worth giving up some formal self-ownership for the sake of substantive self-determination.

To summarize this section, I have argued that Rawlsian redistribution is compatible with formal self-ownership, and that it does a better job than libertarianism in promoting fairly substantive self-ownership. I have also argued that formal self-ownership is a red herring, for substantive self-determination is the more fundamental value. But there is a deeper problem with Nozick's self-ownership argument. Nozick has not adequately confronted Rawls's claim that people do not have a legitimate claim to the rewards of the exercise of their undeserved talents. I have tried to show that we can get a Rawlsian distributive scheme even without denying self-ownership, since redistribution could arise from the requirements of a fair theory of access to external resources. But I still think that Rawls's denial of self-ownership was perfectly sound. I think that we can treat people's talents as part of their circumstances, and hence as possible grounds, in and of themselves, for compensation. People have rights to the possession and exercise of their talents, but the disadvantaged may also have rights to some compensation for their disadvantage. It is wrong for people to suffer from undeserved inequalities in circumstances, and the disadvantaged have direct claims on the more fortunate, quite independently of the question of access to external resources. As I said in discussing his Wilt Chamberlain example, Nozick has not given us any reason to reject that Rawlsian intuition.

3. LIBERTARIANISM AS MUTUAL ADVANTAGE

Many libertarians acknowledge that Nozick's argument fails. The problem, they say, is not with Nozick's conclusions, but with his attempt to defend them by appeal to Kant's egalitarian idea of treating people as ends in themselves. If we start with the idea that each person matters equally, then justice will require something other than Nozickian self-ownership. But, they claim, that just shows that libertarianism is not properly viewed as a theory of treating people as equals. What then is it a theory of? There are two main possibilities: in this section, I will consider libertarianism as a theory of mutual advantage; in the next section, I will consider it as a theory of freedom.

Mutual advantage theories of libertarianism are often presented in contractarian terms. This can be confusing, since liberal egalitarian theories have also been presented in contractarian terms, and the shared use of the contract device can obscure the fundamental differences between them. Before evaluating the mutual advantage defence of libertarianism, therefore, I will lay out some of the differences between the Rawlsian and mutual advantage versions of contractarianism.

For Rawls, the contract device is tied to our 'natural duty of justice'. We have a natural duty to treat others fairly, for they are 'self-originating sources of valid claims'. People matter, from the moral point of view, not because they can harm or benefit us, but because they are 'ends in themselves' (Rawls 1971: 179–80), and so are entitled to equal consideration. This is a 'natural' duty because it is not derived from consent or mutual advantage, but simply owed to persons as such (Rawls 1971: 115–16). The contract device helps us determine the content of this natural duty, for it requires that each party take into consideration the needs of others 'as free and equal beings'. To ensure that the contract gives equal consideration to each of the contractors, Rawls's original position abstracts from differences in talent and strength that might create unequal bargaining power. By removing these arbitrary differences, the contract device 'substitutes a moral equality for physical inequality' (Diggs 1981: 277), and thereby 'represents equality between human beings as moral persons' (Rawls 1971: 19). For Rawls, then, the contract is a useful device for determining the content of our natural duty of justice, because it properly represents our moral equality (ch. 3, s. 3 above).

Mutual advantage theorists also use a contract device, but for opposite reasons. For them, there are no natural duties or self-originating moral claims. There is no moral equality underneath our natural physical inequality. The modern world-view, they say, rules out the traditional idea that people and actions have any inherent moral status. What people take to be objective moral values are just the subjective preferences of individuals (Buchanan 1975: 1; Gauthier 1986: 55–9; Narveson 1988: 110–21).

So there is nothing naturally 'right' or 'wrong' about one's actions, even if they involve harming others. However, while there is nothing inherently wrong in harming you, I would be better off by refraining from doing so if every other person refrains from harming me. Adopting a convention against injury is mutually advantageous—we do not have to waste resources defending our own person and property, and it enables us to enter into stable co-operation. It may be in our short-term self-interest to violate such an agreement on occasion, but acting on short-term self-interest makes mutual co-operation and constraint unstable, and thereby harms our long-term self-interest (it eventually leads to Hobbes's 'war of all against all'). While injury is not inherently wrong, each person gains in the long run by accepting conventions that define it as 'wrong' and 'unjust'.

The content of such conventions will be the subject of bargaining—each person will want the convention to protect their own interests as much as possible while constraining them as little as possible. While conventions are not really contracts, we can view this bargaining over mutually advantageous conventions as the process by which a community establishes its 'social contract'. While this contract, unlike Rawls's, is not an elaboration of our traditional notions of moral and political obligation, it will include some of the constraints that Rawls and others take to be 'natural duties'—for example, the duty not to steal, or the duty to share the benefits of co-operation fairly amongst the contributors. Mutually advantageous conventions occupy some of the place of traditional morality, and, for that reason, can be seen as providing a 'moral' code, even though it is 'generated as a rational constraint from the non-moral premises of rational choice' (Gauthier 1986: 4).

This sort of theory is aptly described by David Gauthier, its best-known proponent, as 'moral artifice', for it is an artificial way of constraining what people are naturally entitled to do. But while the resulting constraints partially overlap with our traditional moral duties, the overlap is far from complete. Whether it is advantageous to follow a particular convention depends on one's preferences and powers. Those who are strong and talented will do better than those who are weak and infirm, since they have much greater bargaining power. The infirm produce little of benefit to others, and what little they do produce may be simply expropriated by others without fear of retaliation. Since there is little to gain from co-operation with the infirm, and nothing to fear from retaliation, the strong do not gain from accepting conventions which recognize or protect the interests of the infirm. This is precisely what Rawls objected to in traditional state-of-nature arguments—they allow differences in bargaining power that should be irrelevant when determining principles of justice. But Gauthier is using the contract device to determine principles of mutual advantage, and differences in bargaining power are central to that question. The resulting conventions will accord rights to various people, but since these rights depend on one's bargaining power, mutual advantage contractarianism does 'not afford each individual an inherent moral status in relation to her fellows' (Gauthier 1986: 222).

It is hard to exaggerate the difference between these two versions of contractarianism. Rawls uses the device of a contract to develop our traditional notions of moral obligation, whereas Gauthier uses it to replace them; Rawls uses the idea of the contract to express the inherent moral standing of persons, whereas Gauthier uses it to generate an artificial moral standing; Rawls uses the device of the contract to negate differences in bargaining power, whereas Gauthier uses it to reflect them. In both premises and conclusions, these two strands of contract theory are, morally speaking, a world apart.

I will question the plausibility of the mutual advantage approach momentarily. But, even if we accept it, how does it justify a libertarian regime in which each person has unfettered freedom of individual contract over her self and her holdings? It cannot, of course, yield self-ownership as a natural right. As Gauthier says, mutual advantage theories do not offer people an 'inherent moral status', and if there are no natural duties to respect others, then obviously there is no natural duty to respect their self-ownership, and hence no duty to treat them in ways they would voluntarily consent or contract to. But libertarians argue that respecting self-ownership is mutually advantageous—it is in each person's interest to accord self-ownership rights to others, and not try to coerce them into promoting our good, so long as they reciprocate. The costs of coercing others are too high, and the pay-offs too low, to be worth the risk of being coerced oneself. Mutual advantage does not, however, justify any further rights—rights to a certain share of resources under Rawls's difference principle, for example. The poor would gain from such a right, but the rich have an interest in protecting their resources, and the poor lack sufficient power to take the resources, or to make the costs of protection exceed its benefits. Mutual advantage yields libertarianism, therefore, because everyone has both the interest and the ability to insist on self-ownership, but those who have an interest in redistribution do not have the ability to insist on it (Harman 1985: 321–2; cf. Barry 1986: ch. 5).

Does mutual advantage justify granting each person rights of self-ownership? Since people lack inherent moral status, whether one has the unfettered right of contract over one's talents and holdings depends on whether one has the power to defend one's talents and holdings against coercion by others. Mutual advantage libertarians claim that everyone does, in fact, have this power. They claim that humans are by nature equal, not in Rawls's sense of sharing a fundamental equality of natural right—rather, equality of rights 'is derivative from a fundamental *factual* equality of condition, in fact an equal vulnerability to the invasions of others' (Lessnoff 1986: 107). People are, by nature, more or less equal in their ability to harm others and their vulnerability to being harmed—and this factual equality grounds equal respect for self-ownership.

But this is unrealistic. Many people lack the power to defend themselves, and so cannot claim the right of self-ownership on mutual advantage grounds. As James Buchanan says, 'if personal differences are sufficiently great', then the strong may have the capacity to 'eliminate' the weak, or perhaps to seize any goods produced by the weak, and thereby set up 'something similar to the slave contract' (Buchanan 1975: 59–60). These are not abstract possibilities—personal differences are that great. It is an inescapable consequence of mutual advantage theories that the congenitally infirm 'fall beyond the pale' of justice (Gauthier 1986: 268), as do young children since 'there is little the child can do to retaliate against those jeopardizing its well-being' (Lomasky 1987: 161; Grice 1967: 147–8).

It is doubtful that many mutual advantage theorists really believe in this assumption of a natural equality in bargaining power. Their claim in the end is not that people are in fact equals by nature, but rather that justice is only possible in so far as this is so. By nature, everyone is entitled to use whatever means are available to them, and the only way moral constraints will arise is if people are more or less equal in their powers and vulnerabilities. For only then does each person gain more from the protection of their own person and property than they lose by refraining from using other people's bodies or resources. Natural equality is not sufficient, however, for people of similar physical capacities may find themselves with radically unequal technological capacities, and 'those with a more advanced technology are frequently in a position to dictate the terms of interaction to their fellows' (Gauthier 1986: 231; Hampton 1986: 255). Indeed, technology may get us to the point where, as Hobbes put it, there is a 'power irresistable' on earth, and for Hobbes and his contemporary followers, such power 'justifieth all actions really and properly, in whomsoever it is found'. No one could claim rights of self-ownership against such power.[6]

Mutual advantage, therefore, subordinates individual self-ownership to the power of others. This is why Nozick made self-ownership a matter of our natural rights. Coercing others is wrong for Nozick, not because it is too costly for the coercer, but because people are ends in themselves, and coercion violates people's inherent moral status by treating them as a means. Nozick's defence of libertarianism, therefore, relies precisely on the premiss that Gauthier denies—namely, that people have inherent moral status. But neither approach actually yields libertarianism. Nozick's approach explains why everyone has equal rights, regardless of their bargaining power, but cannot explain why people's rights do not include some claim on social resources. Gauthier's approach explains why the vulnerable and weak do not have a claim on resources, but cannot explain why they have an equal claim to self-ownership, despite their unequal bargaining power. Treating people as ends in themselves requires more than (or other than) respecting their self-ownership (*contra* Nozick); treating people according to mutual advantage often requires less than respect for self-ownership (*contra* Gauthier).

Let us assume, however, that mutual advantage does lead to libertarianism. Perhaps Lomasky is right that it costs too much to determine who one can enslave and who one must treat as an equal, so that the strong would agree to conventions that accord self-ownership to even the weakest person (Lomasky 1987: 76–7). How would this constitute a defence of libertarianism? On our everyday view, mutually advantageous activities are only legitimate if they respect the rights of others (including the rights of those too weak to defend their interests). It may not be advantageous for the strong to refrain from killing or enslaving the weak, but the weak have prior claims of justice against the strong. To deny this is 'a hollow mockery of the idea of justice—adding insult to injury. Justice is normally thought of not as ceasing to be relevant in conditions of extreme inequality in power but, rather, as being especially relevant in such conditions' (Barry 1989: 163). Exploiting the defenceless is, on our everyday view, the worst injustice, whereas mutual advantage theorists say we have no obligations at all to the defenceless.

This appeal to everyday morality begs the question, since the whole point of the mutual advantage approach is that there are no natural duties to others—it challenges those who believe there is 'a real moral difference between right and wrong which all men [have] a duty to respect' (Gough 1957: 118). To say that Gauthier ignores our duty to protect the vulnerable is not to give an argument against his theory, for the existence of such duties is the very issue in question. But, precisely because it abandons the idea that people have inherent moral status, the mutual advantage approach is not an alternative account of justice, but rather an alternative to justice. While mutual advantage may generate just outcomes under conditions of natural and technological equality, it licenses exploitation wherever 'personal differences are sufficiently great', and there are no grounds within the theory to prefer justice to exploitation. If people act justly, it is not because they see justice as a value, but only because they lack 'power irresistable' and so must settle for justice. From the point of view of everyday morality, therefore, mutual advantage contractarianism may provide a useful analysis of rational self-interest or of *realpolitik*, 'but why we should regard it as a method of moral justification remains utterly mysterious' (Sumner 1987: 158; cf. Barry 1989: 284). As Rawls says, 'to each according to his threat advantage' simply does not count as a conception of justice (Rawls 1971: 134).

None of this will perturb the mutual advantage theorist. If one rejects the idea that people or actions have inherent moral status, then moral constraints must be artificial, not natural, resting on mutually advantageous conventions. And if mutually advantageous conventions conflict with everyday morality, then 'so much the worse for morality' (Morris 1988: 120). Mutual advantage may be the best we can hope for in a world without natural duties or objective moral values.

The mutual advantage approach will be attractive to those who share its scepticism about moral claims. Most political philosophy in the Western tradition, however, shares the opposite view that there are obligation-generating rights and wrongs which all persons have a duty to respect. And, in my view, this is a legitimate assumption. It is true that our claims about natural duties are not observable or testable, but different kinds of objectivity apply to different areas of knowledge, and there is no reason to expect or desire that moral duties have the same kind of objectivity as the physical sciences. As Nagel says, '*if* any values are objective, they are objective *values*, not objective anything else' (Nagel 1980: 98).[7]

This is a difficult question, and some people will remain sceptical about the existence of moral duties. If so, then mutual advantage may be all we have with which to construct social rules. But none of this helps the libertarian, for mutually advantageous conventions may often be non-libertarian. Some people will have the ability to coerce others, violating their self-ownership, and some people will have the ability to take others' property, violating their property-ownership. Mutual advantage, therefore, provides only a very limited defence of property-rights, and what little defence it does provide is not a recognizably moral defence. Most libertarians would prefer to defend their commitment to property-rights in the language of justice, not power. Nozick's appeal to equality did not work, but we have yet to examine the appeal to freedom.

4. LIBERTARIANISM AS LIBERTY

Some people argue that libertarianism is not a theory of equality or mutual advantage. Rather, as the name suggests, it is a theory of liberty. On this view, equality and liberty are rivals for our moral allegiance, and what defines libertarianism is precisely its avowal of liberty as a foundational moral premiss, and its refusal to compromise liberty with equality (unlike the welfare-state liberal).

This is not a plausible interpretation of Nozick's theory. Nozick does say that we are free, morally speaking, to use our powers as we wish. But this self-ownership is not derived from any principle of liberty. He does not say that freedom comes first, and that, in order to be free, we need self-ownership. He gives us no purchase on the idea of freedom as something prior to self-ownership from which we might derive self-ownership. His view, rather, is that the scope and nature of the freedom we ought to enjoy are a function of our self-ownership.

Other libertarians, however, say that libertarianism is based on a principle of liberty. What does it mean for a theory to be based on a principle of liberty, and how does such a principle serve to defend capitalism? One obvious answer is this:

1. an unrestricted market involves more freedom;
2. freedom is the fundamental value;
3. therefore, the free market is morally required.

But this view, while very common, is a serious confusion. I will argue that both (1) and (2), as standardly presented, are misleading, and that any attempt to defend capitalism by appeal to a principle of liberty will encounter similar problems.

(a) The value of liberty

(i) The role of liberty in egalitarian theories

Let us start with premiss (2), concerning the value of liberty. Before examining the claim that liberty is a fundamental value, it is important to clarify the role of liberty in the theories we have examined so far. I have argued that utilitarianism, liberalism, and Nozick's libertarianism are all egalitarian theories. While liberty is not a fundamental value in these theories, that does not mean that they are unconcerned with liberty. On the contrary, the protection of certain liberties was of great importance in each theory. This is obvious in the case of Nozick, who emphasizes the formal liberties of self-ownership, and Rawls, who assigns lexical priority to the basic civil and political liberties. But it is also true of most utilitarians, like Mill, who felt that utility was maximized by according people the freedom to choose their own way of life.

In deciding which liberties should be protected, egalitarian theorists situate these liberties within an account of equal concern for people's interests. They ask whether a particular liberty promotes people's interests, and state that, if so, then it should be promoted because people's interests should be promoted. For example, if each person has an important interest in the freedom to choose their marital partner, then denying someone that liberty denies her the respect and concern she is entitled to, denies her equal standing as a human being whose well-being is a matter of equal concern. Defending a particular liberty, therefore, involves answering the following two questions:

(a) which liberties are important, which liberties matter, given our account of people's interests?

(b) what distribution of important liberties gives equal consideration to each person's interests?

In other words, egalitarian theorists ask how a particular liberty fits into a theory of people's interests, and then ask how a distribution of that liberty fits into a theory of equal concern for people's interests. In Rawls's case, for example, we ask what scheme of liberties would be chosen from a contracting position that represents impartial concern for people's interests. In this way, particular liberties can come to play an important role in egalitarian theories. I will call this the 'Rawlsian approach' to assessing liberties.

Mutual advantage theories assess liberty in a similar way. Like Rawls, they ask which particular liberties promote people's interests, and then ask what distribution of these liberties follows from a proper weighing of people's interests. The only difference is that in mutual advantage theories people's interests are weighed according to their bargaining power, not according to impartial concern. In Gauthier's case, for example, we ask what scheme of liberties would be agreed to by contractors negotiating for mutual advantage on the basis of their interests.

As we have seen, many libertarians defend their preferred liberties (e.g. the freedom to exercise one's talents in the market) in one of these two ways. Indeed, some of the libertarians who say that their theory is 'liberty-based' also defend their preferred liberties in terms of consideration for people's interests, weighed according to the criteria of equality or mutual advantage. They call that a liberty-based argument, to emphasize their belief that our essential interest is an interest in certain kinds of liberty, but this new label does not affect the underlying argument. And, regardless of the label, assessing liberties in terms of either equality or mutual advantage will not yield libertarianism, for reasons I have discussed.

Can the defence of libertarianism be liberty-based in a way that is genuinely different from a defence based on equality or mutual advantage? What would it mean for libertarians to defend their preferred freedoms by appealing to a principle of liberty? There are two possibilities. One principle of liberty is that freedom should be maximized in society. Libertarians who appeal to this principle defend their preferred liberties by claiming that the recognition of these particular liberties maximizes freedom in society. The second principle of liberty is that people have a right to the most extensive liberty compatible with a like liberty for all. Libertarians who appeal to this principle defend their preferred liberties by claiming that recognizing these particular liberties increases each person's overall freedom. I will argue that the first principle is absurd, and has no attraction to anyone, including libertarians; and the second principle is either a confused way of restating the egalitarian argument, or it rests on an indeterminate and unattractive conception of freedom. Moreover, even if we accept the absurd or unattractive interpretations of the principle of liberty, they still will not defend libertarianism.

(ii) Teleological liberty

The first principle of liberty says that we should aim to maximize the amount of freedom in society. If freedom is the ultimate value, why not aim to have as much of it as possible? This, of course, is the way teleological utilitarians argue for the maximization of utility, so I will call this the 'teleological' liberty principle. But, as we saw in chapter 2, this sort of theory loses touch with our most basic understanding of morality. Because teleological theories take concern for the good (e.g. freedom or utility) as fundamental, and concern for people as derivative, promoting the good becomes detached from promoting people's interests. For example, we could increase the amount of freedom in society by increasing the number of people, even if each person's freedom is unchanged. Yet no one thinks that a more populous country is, for that reason alone, more free in any morally relevant sense.

Indeed, it may be possible to promote the good by sacrificing people. For example, a teleological principle could require that we coerce people to bear and raise children and thereby increase the population. This deprives existing people of a freedom, but the result would increase the overall amount of freedom, since the many freedoms of the new population outweigh the loss of one freedom amongst the earlier population. The principle could also justify unequally distributing liberties. If five people enslave me, there is no reason to assume that the loss of my freedom outweighs the increased freedom of the five slave-owners. They may gain more options or choices collectively from the freedom to dispose of my labour than I lose (assuming that it is possible to measure such things—see s. 4*a*(iii) below). No libertarian supports such policies, for they violate fundamental rights.

Whatever libertarians mean by saying their theory is liberty-based, it cannot be this. On a teleological liberty principle, the goal is not to respect people, for whom certain liberties are needed or wanted, but to respect liberty, for which certain people may or may not be useful contributors. Libertarianism, however, respects people first, and respects liberty as one component of respect for people.

So libertarians do not favour maximizing freedom in society. Yet it is a natural interpretation of the claim that freedom is the fundamental value, and it is encouraged by the libertarian's rhetorical rejection of equality. Libertarians believe in equal rights of self-ownership, but many of them do not want to defend this by appeal to any principle of equality. They try to find a liberty-based reason for equally distributing liberties. Thus some libertarians say that they favour equal liberties because they believe in freedom, and since each individual can be free, each individual should be free.[8] But if this really were the explanation of the libertarian commitment to equal liberty, then they should increase the population, since future people too can be free. Libertarian attempts to defend equal liberty by appeal to liberty, rather than equality, pull them in the direction of a teleological principle of liberty, for it is the one principle of liberty that makes no appeal to equality. But this misdescribes their real commitments. Libertarians reject increasing the overall amount of freedom through increasing the population, and they reject it for the same reason they reject increasing the overall amount of freedom by unequally distributing liberties— namely, their theory is equality-based. As Peter Jones puts it, 'to prefer equal liberty to unequal liberty is to prefer equality to inequality, rather than freedom to unfreedom' (Jones 1982: 233). So long as libertarians are committed to equal liberty for each person, they are adopting an equality-based theory. Their attempt to deny this by putting everything in terms of liberty not only confuses the issue, but threatens to undermine their own commitment to self-ownership.

(iii) Neutral liberty

The second, and more promising, candidate for a foundational principle of liberty says that each person is entitled to the most extensive liberty compatible with a like liberty for all. I will call this the 'greatest equal liberty' principle. This principle works within the general framework of an egalitarian theory, since now equal liberty cannot be sacrificed for a greater overall liberty, but it is importantly different from the Rawlsian approach (s. 4*a*(i) above). The Rawlsian approach assessed particular liberties by asking how they promote our interests. The greatest equal liberty approach assesses particular liberties by asking how much freedom they give us, on the assumption that we have an interest in freedom as such, in maximizing our overall freedom. Both approaches connect the value of particular liberties to an account of our interests. But the Rawlsian approach did not say that we have an interest in freedom as such, or that our interest in any particular liberty corresponds to how much freedom it contains, or that it even makes sense to compare the amount of freedom contained in different liberties. Different liberties promote different interests for many different reasons, and there is no reason to assume that the liberties which are most valuable to us are the ones with the most freedom. The greatest equal liberty approach, however, says that the value of any particular liberty just is how much freedom it contains, for our interest in particular liberties stems from our interest in freedom as such. Unlike the Rawlsian approach, judgements of the value of different liberties require, and are derived from, judgements of greater or lesser freedom.

If libertarianism appeals to this greatest equal liberty principle, then it is not a 'liberty-based' theory in the strict sense, for (unlike a teleological liberty-based theory) rights to liberty are derived from the claims of people to equal consideration. But it is liberty-based in a looser sense, for (unlike the Rawlsian approach to liberty) it derives judgements of the value of particular liberties from judgements of greater or lesser freedom. Can the libertarian defend his preferred liberties by appeal to the greatest equal liberty principle? Before we can answer that question, we need some way of measuring freedom, so that we can determine whether the free market, for example, maximizes each individual's freedom.

In order to measure freedom, we need to define it. There are many definitions of freedom in the literature, but they can be grouped, for the moment, into two camps (further subdivisions will be considered later). The 'Lockean' camp defines freedom in terms of the exercise of our rights. Whether or not a restriction decreases our freedom depends on whether or not we had a right to do the restricted thing. For example, Lockeans say that preventing someone from stealing is not a restriction on their liberty, since they had no right to steal. This is a 'moralized' definition of liberty, since it presupposes a prior theory of rights. The 'Spencerian' camp, on the other hand, defines liberty in a non-moralized way—as the presence of options or choices, for example—without assuming that we have a right to exercise those options. Spencerians then assign rights so as to maximize each individual's freedom, compatible with a like freedom for all. Hence whether people have a right to appropriate previously unowned natural resources depends on whether that right increases or decreases each person's freedom (cf. Sterba 1988: 11–15).

If the greatest equal liberty principle is to be foundational, then the Lockean definition is excluded. If we are trying to derive rights from judgements of greater or lesser liberty, our definition of liberty cannot presuppose some principle of rights. Libertarians who appeal to the greatest equal liberty principle believe that whether we have a right to appropriate unowned resources, for example, depends on whether that right increases each person's freedom. But on a Lockean definition of freedom, we first need to know whether people have a right to appropriate unowned resources in order to know whether a restriction on appropriation is a restriction on their freedom. So the libertarian defence of capitalism, if it is to be liberty-based, must use a Spencerian definition of freedom. (The Spencerian definition is preferable anyway, for the Lockean definition is at odds with our everyday usage of 'freedom'. In everyday conversation, we say that a prisoner is deprived of her freedom, even if her incarceration is legitimate.)

Within the Spencerian camp, there are two main proposals for a non-moralized definition of liberty. One is a 'neutral' definition, the other 'purposive'. Each definition purports to provide a criterion for determining whether a particular liberty increases someone's overall freedom, as is required by the principle of greatest equal liberty. I will argue that using the first definition is unattractive, and yields indeterminate results, while the second is simply a confused way of restating the Rawlsian approach to assessing liberties.

On the 'neutral' view, we are free in so far as no one prevents us from acting on our desires. This is a non-moralized definition since it does not presuppose that we have a right to act on these desires. Using this definition we may be able to make comparative judgements about the quantity of one's freedom. One can be more or less free, on this definition, since one can be free to act on some but not other desires. If we can make such quantitative judgements about the amount of freedom provided by different rights, then we can determine which rights are most valuable. If the principle of greatest equal liberty employs this definition of freedom, then each person is entitled to the greatest amount of neutral freedom compatible with a like freedom for all.

But does this in fact provide a standard for assessing the value of different liberties? There are two main problems. Firstly, our intuitive judgements about the value of different liberties are not in fact based on quantitative judgements of neutral freedom. Hence such judgements lead to counter-intuitive results. Secondly, the required quantitative judgements of neutral freedom may be impossible to make.

Do quantitative judgements of neutral freedom underlie our everyday assessment of the value of different liberties? Compare the inhabitants of London with citizens of an underdeveloped Communist country like Albania. We normally think of the average Londoner as better off in terms of freedom. After all, she has the right to vote, and practise her religion, as well as other civil and democratic liberties. The Albanian lacks these. On the other hand, Albania does not have many traffic lights, and those people who own cars face few if any legal restrictions on where or how they drive. The fact that Albania has fewer traffic restrictions does not change our sense that Albanians are worse off, in terms of freedom. But can we explain that fact by appealing to a quantitative judgement of neutral freedom?

If freedom can be neutrally quantified, so that we can measure the number of times each day that traffic lights legally prevent

Londoners from acting in a certain way, there is no reason to say that these will outnumber the times that Albanians are legally prevented from practising religion in public. As Charles Taylor (from whom I have taken the example) puts it, 'only a minority of Londoners practise some religion in public places, but all have to negotiate their way through traffic. Those who do practise a religion generally do so on one day of the week, while they are held up at traffic lights every day. In sheer quantitative terms, the number of acts restricted by traffic lights must be greater than that restricted by a ban on public religious practice' (Taylor 1985: 219).

Why do we not accept Taylor's 'diabolical defence' of Albanian freedom—why do we think that the Londoner is better off in terms of freedom? Because restrictions on civil and political liberty are more important than restrictions on traffic mobility. They are more important, not because they involve *more freedom*, neutrally defined, but because they involve *more important freedoms*. They are more important because, for example, they allow us to have greater control over the central projects in our lives, and so give us a greater degree of self-determination, in a way that traffic freedoms do not, whether or not they involve a smaller quantity of neutral freedom.

The neutral view of liberty says that each neutral freedom is as important as any other. But when we think about the value of different liberties in relation to people's interests, we see that some liberties are more important than others, and indeed some liberties are without any real value—e.g. the freedom to libel others (Hart 1975: 245). Our theory must be able to explain the distinctions we make amongst different kinds of liberty.

The problems for neutral freedom go still deeper. The required judgements of greater or lesser freedom may be impossible to make, for there is no scale on which to measure quantities of neutral freedom. I said earlier that if we could count the number of free acts restricted by traffic laws and political censorship, traffic laws would probably restrict more free acts. But the idea of a 'free act' is an elusive one. How many free acts are involved in the simple waving of a hand? If a country outlaws such waving, how many acts has it forbidden? How do we compare that to a restriction on religious ceremonies? In each case, we could, with equally much or little justification, say that the laws have outlawed one act (waving a hand, celebrating religious belief), or that they have outlawed an infinite number of acts, which could have been performed an infinite number of times. But the principle of greatest equal liberty requires the ability to discriminate between these two cases. We need to be able to say, for example, that denying religious ceremonies takes away five units of free acts, whereas denying waving of one's hand takes away three. But how we could go about making such judgements is quite mysterious. As O'Neill puts it, 'We can, if we want to, take any liberty—e.g. the liberty to seek public office or the liberty to form a family—and divide it up into however many component liberties we find useful to distinguish— or for that matter into more than we find it useful to distinguish' (O'Neill 1980: 50). There is no non-arbitrary way of dividing up the world into actions and possible actions which would allow us to say that more neutral freedom is involved in denying free traffic movement than in denying free speech. (The one exception involves comparing two essentially identical sets of rights, where the second set contains all the neutral freedoms in the first set, plus at least one more free act—see Arneson 1985: 442–5.)

Traffic laws and political oppression both restrict free acts. But any attempt to weigh the two on a single scale of neutral freedom, based on some individuation and measurement of free acts, is implausible. There may be such a scale, but those libertarians who endorse a neutral version of the greatest equal liberty principle have not made many strenuous attempts to develop such a scale (for one attempt, see Steiner 1983). It is interesting that people who say that quantitative judgements of neutral freedom are not only possible but fundamental have not actually tried to show how one can measure the amount of neutral freedom in religious liberties as opposed to traffic liberties.

How then do libertarians who endorse the neutral version of the greatest equal liberty principle reach any determinate conclusions? As I will show in section 4b below, they often defend their preferred liberties by simply ignoring the loss of neutral freedom involved in libertarian policies, or by invoking *ad hoc* criteria for preferring one set of liberties to another.

(iv) Purposive liberty

Our most valued liberties (the ones that make us attracted to a principle of greatest equal liberty) do not seem to involve the greatest neutral freedom. The obvious move, for advocates of the greatest equal liberty principle, is to adopt a 'purposive' definition of liberty. On such a definition, the amount of freedom contained in a particular liberty depends on how important that liberty is to us, given our interests and purposes. As Taylor puts it, 'Freedom is important to us because we are purposive beings. But then there must be distinctions in the significance of different kinds of freedom based on the distinction in the significance of different purposes' (Taylor 1985: 219). For example, religious liberty gives us more freedom than traffic liberty because it serves more important interests, even if it does not contain quantitatively more neutral freedom.

A purposive definition of freedom requires some standard for assessing the importance of a liberty, in order to measure the amount of freedom it contains. There are two basic standards—a 'subjective' standard says the value of a particular liberty depends on how much an individual desires it; an 'objective' standard says that certain liberties are important whether or not a particular person desires them. The latter is often thought to be preferable because it avoids the potential problem of the 'contented slave' who does not desire legal rights, and hence, on a subjective standard, does not lack any important freedoms.

On either view, we assess someone's freedom by determining how valuable (subjectively or objectively) her specific liberties are. Those liberties that are more highly valued contain, for that reason, more purposive freedom. On the purposive version of the greatest equal liberty principle, therefore, each person is entitled to the greatest possible amount of purposive liberty compatible with a like liberty for all. Like the Rawlsian approach to assessing liberties, this allows for qualitative judgements of the value of particular liberties, but it differs from the Rawlsian approach in supposing that these liberties must be assessed in terms of a single scale of freedom.

This is more attractive than the neutral version, for it corresponds with our everyday view that some neutral freedoms are more valuable than others. The problem, however, is that the whole language of greater and lesser freedom is no longer doing any work in the argument. The purposive version of the greatest equal liberty principle is in fact just a confused way of restating the Rawlsian approach. It seems to differ in saying that the reason we are entitled to important liberties is that we are entitled to the greatest amount of equal liberty, a step that is absent in the Rawlsian approach. But that step does no work in the argument, and indeed simply confuses the real issues.

The principle of greatest equal liberty provides the following argument for protecting a particular liberty:

1. Each person's interests matter and matter equally.
2. People have an interest in the greatest amount of freedom.
3. Therefore, people should have the greatest amount of freedom, consistent with the equal freedom of others.
4. The liberty to x is important, given our interests.
5. Therefore, the liberty to x increases our freedom.
6. Therefore, each person ought (*ceteris paribus*) to have the right to x, consistent with everyone else's right to x.

Contrast that with the Rawlsian argument:

1. Each person's interests matter and matter equally.
4. The liberty to x is important, given our interests.
6. Therefore, each person ought (*ceteris paribus*) to have the right to x, consistent with everyone else's right to x.

The first argument is a needlessly complex way of stating the second argument. The step from (4) to (5) adds nothing (and, as a result, steps (2) and (3) also add nothing). Libertarians, on this view, say that because a particular liberty is important, therefore it increases our freedom, and we should have as much freedom as possible. But, in fact, the argument for the liberty is completed with the assessment of its importance.

Consider Loevinsohn's theory of measuring freedom, which uses a subjective standard for measuring purposive freedom. He says that 'when force or the threat of penalties is used to prevent someone from pursuing some possible course of action, the degree to which his liberty is thereby curtailed depends . . . on how important the course of action in question is to him' (Loevinsohn 1977: 232; cf. Arneson 1985: 428). Hence the more I desire a liberty, the more freedom it provides me. If I desire religious liberty more than traffic liberty, because it promotes important spiritual interests, then it gives me more freedom than traffic liberty. But Loevinsohn does not explain what is gained by shifting from the language of 'a more desired liberty' to 'more freedom'. This redescription (the move from (4) to (5) in the above argument) adds nothing, and so the principle of greatest equal liberty ((2) and (3) above) is doing no work. I do not mean that it is impossible to redescribe more desired liberties as more extensive freedom—we can use words however we like. But the fact that we can redescribe them in this way does not mean that we have said anything of moral significance, or that we have found a distinctly liberty-based way of assessing the value of particular liberties.

The greatest equal liberty premiss is not only unnecessary, it is confusing, for a number of reasons. For one thing, it falsely suggests that we have just one interest in liberty. Saying that we evaluate different liberties in terms of how much purposive freedom they provide suggests that these different liberties are important to us for the same reason, that they all promote the same interest. But in fact different liberties promote different interests in different ways. Religious liberties are important for self-determination—i.e. for acting on my deepest values and beliefs. Democratic liberties serve a more symbolic interest—denying me the vote is an assault on my dignity, but may have no effect on my ability to pursue my goals. Some economic liberties have a purely instrumental value—I may desire free trade between countries because it reduces the price of consumer goods, but I would support restrictions on international trade if doing so lowered prices. I do not desire these different liberties for the same reason, and the strength of my desire is not based on the extent to which they promote a single interest.[9] Again, it is possible to redescribe these different interests as an interest in a more extensive purposive freedom. One can say that desiring a particular liberty (for whatever reason) just is desiring to have more extensive freedom. But it is needlessly confusing.

Moreover, talking about our interest in more extensive freedom, as opposed to our different interests in different liberties, obscures the relationship between freedom and other values. Whatever interest we have in a particular liberty—be it intrinsic or instrumental, symbolic or substantive—it is likely that we have the same interest in other things. For example, if the freedom to vote is important for its effect on our dignity, then anything else that promotes our dignity is also important (e.g. meeting basic needs, or preventing libel), and it is important for the very same reason. The defender of purposive freedom says that our concern is with important liberties, not just any old neutral liberty. But if we look at what makes liberties important to us, then freedom no longer systematically competes with other values like dignity, or material security, or autonomy, for these often are the very values which make particular liberties important. Describing more important liberties as more extensive freedom, however, invites this false contrast, for it pretends that the importance of particular liberties stems from the amount of freedom they contain.

So neither version of the greatest equal liberty principle offers a viable alternative to the Rawlsian approach to assessing liberties. It is worth noting that Rawls himself once endorsed a right to the most extensive equal liberty, and it was only in the final version of his theory that he adopted what I have called the Rawlsian approach. He now defends a principle of equal rights to 'basic liberties', while disavowing any claims about the possibility, or significance, of measurements of overall freedom (Rawls 1982*a*: 5–6; Hart 1975: 233–9). He recognized that in determining which are the basic liberties, we do not ask which liberties maximize our possession of a single commodity called 'freedom'. Of course, the claim that people are maximally free is often 'merely elliptical for the claim that they are free in every important respect, or in most important respects' (MacCallum 1967: 329). But as Rawls recognized, once we say this, then the principle of greatest equal liberty does no work. For the reason it is important to be free in a particular respect is not the amount of freedom it provides, but the importance of the various interests it serves. As Dworkin puts it,

If we have a right to basic liberties not because they are cases in which the commodity of liberty is somehow especially at stake, but because an assault on basic liberties injures us or demeans us in some way that goes beyond its impact on liberty, then what we have a right to is not liberty at all, but to the values or interests or standing that this particular constraint defeats. (Dworkin 1977: 271)

In making liberty-claims, therefore, we are entitled, not to the greatest equal amount of this single commodity of freedom, but to equal consideration for the interests that make particular liberties important.

(b) Freedom and capitalism

Many libertarians defend property-rights on the basis of a principle of liberty. I have considered three possible definitions of liberty that could be used in this defence. Lockean definitions will not work, because they presuppose a theory of rights. The neutral definition will not work, because quantitative measurements of neutral freedom lead to indeterminate or implausible results. And the purposive definition simply obscures the real basis of our assessment of the value of liberties. If those claims are correct, we are left with a puzzle. Why have so many people thought a principle of liberty helps defend capitalism? I will try to show that standard libertarian arguments rely on inconsistent definitions of freedom.

Libertarians often equate capitalism with the absence of restrictions on freedom. Anthony Flew, for example, defines libertarianism as 'opposed to any social and legal constraints on individual freedom' (Flew 1979: 188; cf. Rothbard 1982: v). He contrasts this with liberal egalitarians and socialists who favour government restrictions on the free market. Flew thus identifies capitalism with the absence of restrictions on freedom. Many of those who favour constraining the market agree that they are thereby restricting liberty. Their endorsement of welfare-state capitalism is said to be a compromise between freedom and equality, where freedom is understood as the free market, and equality as welfare-state restrictions on the market. This equation of capitalism and freedom is part of the everyday picture of the political landscape.

Does the free market involve more freedom? It depends on how we define freedom. Flew seems to be assuming a neutral definition of freedom. By eliminating welfare-state redistribution, the free market eliminates some legal constraints on the disposal of one's resources, and thereby creates some neutral freedoms. For example, if government funds a welfare programme by an 80-per-cent tax on inheritance and capital gains, then it prevents people from giving their property to others. Flew does not tell us how much neutral freedom would be gained by removing this tax, but it clearly would allow someone to act in a way they otherwise could not. This expansion of neutral freedom is the most obvious sense in which capitalism increases freedom, but many of these neutral freedoms will also be valuable purposive freedoms, for there are important reasons why people might give their property to others. So capitalism does provide certain neutral and purposive freedoms unavailable under the welfare state.

But we need to be more specific about this increased liberty. Every claim about freedom, to be meaningful, must have a triadic structure—it must be of the form '*x* is free from *y* to do *z*', where *x* specifies the agent, *y* specifies the preventing conditions, and *z* specifies the action. Every freedom claim must have these three elements: it must specify who is free to do what from what obstacle (MacCallum 1967: 314). Flew has told us the last two elements—his claim concerns the freedom to dispose of property without legal constraint. But he has not told us the first—i.e. who has this freedom? As soon as we ask that question, Flew's equation of capitalism with freedom is undermined. For it is the owners of the resource who are made free to dispose of it, while non-owners are deprived of that freedom. Suppose that a large estate you would have inherited (in the absence of an inheritance tax) now becomes a public park, or a low-income housing project (as a result of the tax). The inheritance tax does not *eliminate* the freedom to use the property, rather it *redistributes* that freedom. If you inherit the estate, then you are free to dispose of it as you see fit, but if I use your backyard for my picnic or garden without your permission, then I am breaking the law, and the government will intervene and coercively deprive me of the freedom to continue. On the other hand, my freedom to use and enjoy the property is increased when the welfare state taxes your inheritance to provide me with affordable housing, or a public park. So the free market legally restrains my freedom, while the welfare state increases it. Again, this is most obvious on a neutral definition of freedom, but many of the neutral freedoms I gain from the inheritance tax are also important purposive ones.

That property-rights increase some people's freedom by restrict- ing others' is obvious when we think of the origin of private property. When Amy unilaterally appropriated land that had previously been held in common, Ben was legally deprived of his freedom to use the land. Since private ownership by one person presupposes non-ownership by others, the 'free market' restricts as well as creates liberties, just as welfare-state redistribution both creates and restricts liberties. Hence, as Cohen puts it, 'the sentence "free enterprise constitutes economic liberty" is demonstrably false' (Cohen 1979: 12; cf. Gibbard 1985: 25; Goodin 1988: 312–13).

This undermines an important claim Nozick makes about the superiority of his theory of justice to liberal redistributive theories. He says that Rawls's theory cannot be 'continuously realized without continuous interference with people's lives' (Nozick 1974: 163).

This is because people, left to their own devices, will engage in free exchanges that violate the difference principle, so that preserving the difference principle requires continually intervening in people's exchanges. Nozick claims that his theory avoids continuous interference in people's lives, for it does not require that people's free exchanges conform to a particular pattern, and hence does not require intervening in those exchanges.[10] Unfortunately, the system of exchanges which Nozick protects itself requires continuous interference in people's lives. Left to their own devices, people would freely use the resources around them, and it is only continuous state intervention that prevents them from violating Nozick's principles of justice.

Since property-rights constitute legal restrictions on individual freedom, and since libertarians (according to Flew's definition) claim to be against any such restrictions, one might expect them to demand the abolition of property-rights. But they do not. One might expect them to claim that allowing property-rights creates a larger quantity of freedom than is lost. And indeed some libertarians do make that claim. But it is unclear how we can assess that claim, and even if we could draw up a scale on which property-owners gain more freedom from their rights than the propertyless lose from the resulting restrictions, this maximizing overall liberty claim would not defend libertarianism. It leads to a teleological liberty-based theory that subordinates individual self-ownership to the sum of overall liberty. What libertarians need is the claim that unrestricted property-rights pass the greatest equal liberty test. But then the supposed fact that property-owners gain more liberty than the propertyless lose not only fails to defend libertarianism, it refutes it. What then is the connection between the free market and freedom? Flew's definition implies that the free market does not create any unfreedom, and hence there is no need to weigh the gains against the losses.

How can the free market be seen as an unmitigated increase in freedom? On a non-moralized definition of freedom, private property inherently creates both freedom and non-freedom. If libertarians deny that the free market creates any unfreedom, they must be using a 'Lockean' moralized definition of liberty, one which defines freedom in terms of the exercise of one's rights. My freedom is lessened only when someone prevents me from doing something I have a right to do. If someone has a right to private property, then protecting her property against trespass does not diminish my freedom in any way. Since I had no right to trespass on her property, my freedom is not diminished by the enforcement of property-rights. But once libertarians adopt this moralized definition, the claim that free market increases people's freedom requires a prior argument for the existence of property-rights, an argument which cannot be liberty-based.

To defend the claim that the free market increases freedom, morally defined, libertarians must show that people have a right to property. If people have that right, then respecting the free market increases freedom, and preventing others from using one's property would not count as diminishing their freedom (since no one has a right to trespass). Hence the free market increases freedom, morally defined. But this is not an argument from liberty to property-rights. On the contrary, the liberty claim presupposes the existence of property-rights—property-rights only increase freedom if we have some prior and independent reason to view such rights as morally legitimate.

The claim that the free market increases freedom, therefore, relies on inconsistent definitions of freedom. Libertarians take it as obvious that intervention in private property diminishes freedom. That is true on a Spencerian definition of liberty, and indeed it is the loss of freedom in the non-moralized sense that they point to in defending the claim that capitalism increases freedom. But on a Spencerian definition, it is equally obvious that enforcing property-rights diminishes freedom. To show that the free market increases freedom, the libertarian would have to provide measurements showing that the liberties gained from ownership outweigh the restrictions on liberty created by non-ownership. (This would have to be true for each individual on the greatest equal liberty principle, whereas the teleological principle only requires this to be true for society overall.) But libertarians do not make those measurements. Instead, they argue that property-rights do not create any unfreedom at all, measured by a moralized Lockean definition. This undermines the force of the objection that property-rights diminish freedom, since that objection relied on a non-moralized definition of liberty. However, it also undermines the force of the initial argument libertarians gave to show that property-rights increase freedom, for that too relied on a non-moralized definition. On a moralized definition, it is no longer obvious that unrestricted property rights increase freedom, because it is not obvious that anyone should have unrestricted rights over their property. Indeed, as we have seen, that is a counter-intuitive and implausible claim.

Can any definition of liberty, used consistently, support the claim that libertarianism provides greater equal freedom than a liberal redistributive regime? What if libertarians stick consistently to the neutral definition of liberty, and claim that the free market increases one's overall amount of neutral freedom? Firstly, one must show that the gains in neutral liberty from allowing private property outweigh the losses. Libertarians have not given us any reason to believe that this is true, or that it is possible to carry out the required measurements. Moreover, even if it did increase one's neutral freedom, we would still want to know how important these neutral freedoms are. If our attachment to the free market is only as strong as our attachment to the freedom to libel others, or to run through red lights, then we would not have a very strong defence of capitalism.

What if libertarians adopt the purposive definition, and claim that the free market provides us with the most important liberties? Whether or not unrestricted property-rights promote one's most important purposes depends on whether or not one actually has property. Being free to bequeath property can promote one's most important purposes, but only if one has property to bequeath. So whatever the relationship between property and purposive freedom, the aim of providing the greatest equal freedom suggests an equal distribution of property, not unrestricted capitalism. Nozick denies this, by saying that formal rights of self-ownership are the most important liberties even to those who lack property. But, we have seen, the notion of dignity and agency that Nozick relies on, based on the idea of acting on one's conception of oneself, requires rights over resources as well as one's person. Having independent access to resources is important for our purposes, and hence our purposive freedom, but that argues for liberal equality not libertarianism.

What if libertarians stick to the Lockean definition of liberty, and claim that the free market provides the freedom we have a right to? On a moralized definition, we can only say that respecting a certain liberty increases our freedom if we already know that we have a right to that liberty. And libertarians have given us no plausible argument that we have a right to unrestricted property-ownership. Such a right will not come out of a plausible theory of equality (because it allows undeserved inequalities to have too much influence), nor will it come out of a plausible theory of mutual advantage (because it allows undeserved inequalities to have too little influence). It is difficult to see how any other argument can avoid these seemingly insurmountable objections. But even if we come up with a plausible conception of equality or mutual advantage which includes capitalist property-rights, it is hopelessly confusing then to say that it is an argument about freedom.

None of the three definitions of liberty supports the view that libertarianism increases freedom. Moreover, the failure of these three approaches suggests that the very idea of a liberty-based theory is confused. Our commitment to certain liberties does not derive from any general right to liberty, but from their role in the best theory of equality (or mutual advantage). The question is which specific liberties are most valuable to people, given their essential interests, and which distribution of those liberties is legitimate, given the demands of equality or mutual advantage? The idea of freedom as such, and lesser or greater amounts of it, does no work in political argument.

Scott Gordon objects to this elimination of 'freedom' as a category of political evaluation, and its replacement with the evaluation of specific freedoms: 'If one is driven . . . to greater and greater degrees of specification, freedom as a philosophical and political problem would disappear, obscured altogether by the innumerable specific "freedoms"' (Gordon 1980: 134). But, of course, this is just the point. There is no philosophical and political problem of freedom as such, only the real problem of assessing specific freedoms. Whenever someone says that we should have more freedom, we must ask who ought to be more free to do what from what obstacle? Contrary to Gordon, it is not the specification of these things, but the failure to specify them, that obscures the real issues.[11] Whenever someone tries to defend the free market, or anything else, on the grounds of freedom, we must demand that they specify which people are free to do which sorts of acts—and then ask why those people have a legitimate claim to those liberties—i.e. which interests are promoted by these liberties, and which account of equality or mutual advantage tells us that we ought to attend to those interests in that way. We cannot pre-empt these specific disputes by appealing to any principle or category of freedom as such.

5. THE POLITICS OF LIBERTARIANISM

Libertarianism shares with liberal equality a commitment to the principle of respect for people's choices, but rejects the principle of rectifying unequal circumstances. Taken to the extreme, this is not only intuitively unacceptable, but self-defeating as well, for the failure to rectify disadvantageous circumstances can undermine the very values (e.g. self-determination) that the principle of respect for choices is intended to promote. The libertarian denial that undeserved differences in circumstances give rise to moral claims suggests an almost incomprehensible failure to recognize the profound consequences of such differences.

In practice, however, libertarianism may have a slightly different complexion. Libertarianism gains much of its popularity from a kind of 'slippery-slope' argument which draws attention to the ever-increasing costs of trying to meet the principle of equalizing circumstances. Like Rawls, the libertarian sees the popular conception of equality of opportunity as unstable. If we think social disadvantages should be rectified, then there is no reason not to rectify natural disadvantages. But, libertarians say, while unequal circumstances in principle give rise to legitimate claims, the attempt to implement that principle inevitably leads in practice down a slippery slope to oppressive social intervention, centralized planning, and even human engineering. It leads down the road to serfdom, where the principle of respect for choices gets swallowed up by the requirement to equalize circumstances.

Why might this be? Liberals hope to balance the twin demands of respecting choices and rectifying circumstances. In some cases, this seems unproblematic. The attempt to equalize educational facilities—e.g. to ensure that schools in black neighbourhoods are as good as white schools—does not impinge in an oppressive way on individual choice. Removing well-entrenched inequalities between different social groups requires little intervention in, or even attention to, discrete individual choices. The inequalities are so systematic that no one could suppose that they are traceable to different choices of individuals. But the principle of equalizing circumstances applies to disparities not only between social groups, but also between individuals, and it is less obvious whether those differences are due to choices or circumstances. Consider the problem of effort. In defending the principle of ambition-sensitivity, I used the example of the gardener and the tennis-player, who legitimately come to have differential income due to differential effort. It was important for the success of that example that the two people are similarly situated—i.e. there are no inequalities in skill or education which could prejudice one person's ability to make the relevant effort. But in the real world there is always some difference in people's background which could be said to be the cause of their different choices.

For example, differences in effort are sometimes related to differences in self-respect, which are in turn often related to environmental factors. Some children have more supportive parents or friends, or simply benefit from the contingencies of social life (e.g. not being sick for a test). These different influences will not be obvious, and any serious attempt to establish their presence will be severely invasive. Rawls says that the 'social bases of self-respect' are perhaps the most important primary good (Rawls 1971: 440), but do we want governments measuring how supportive parents are? The situation is different, of course, when it is public institutions which cause differential self-respect. The US Supreme Court struck down segregated education for blacks, even where the facilities were funded equally, because it was perceived as a badge of inferiority, damaging black children's motivation and self-respect. Conversely, some feminists argue that integrated education of boys and girls has negatively affected girls' self-respect, and that single-sex education with equal facilities would better support self-respect. These attempts to equalize the social bases of self-respect can be implemented without undue restriction on respect for individual choice. But, again, the ability to do so stems from the fact that the differences are so readily identifiable at the group level. The differences between whites and blacks, or boys and girls, are so systematic that we do not have to pry or intervene in the details of any particular individual's background or personality. But the situation is more complex when the differences are at a purely individual level.

Moreover, rather than compensate for the effect of unequal circumstances on effort, why not ensure that there are no differential influences on effort, by bringing up children identically? Liberals regard that as an unacceptable restriction on choice. But the libertarian fears it is a logical culmination of the liberal egalitarian commitment to equalizing circumstances. The liberal wants to equalize circumstances in order more fully to respect choices, but, once we move to the level of individual differences and subjective dispositions, the former will swallow the latter.

And why not extend the principle of equalizing circumstances to human engineering, or at least to certain kinds of biological transfers? If one person is born blind and another person is born with two good eyes, why not require the transfer of one good eye to the blind man (Nozick 1974: 206; Flew 1989: 159)? Dworkin points out that there is a difference between changing things so that people are treated as equals, and changing people so that they are, as changed, equal. The principle of equalizing circumstances requires the former, for it is part of the more general requirement that we treat people as equals (Dworkin 1983: 39; Williams 1971: 133–4). That is a valid distinction, but it does not avoid all the problems, for on Dworkin's own theory, people's natural talents are part of their circumstances ('things used in pursuing the good'), not part of the person ('beliefs which define what a good life is about'). So why should eye transfers count as changing people, rather than simply changing their circumstances? Dworkin says that some features of our human embodiment can be both part of the person (in the sense of a constitutive part of our identity) and part of a person's circumstances (a resource). Again that seems sensible. But the lines will not be easy to draw. Where does blood fit in? Would we be changing people if we required healthy people to give blood to haemophiliacs? I do not think so. But what then about kidneys? Like blood, the presence of a second kidney is not an important part of our self-identity, but we are reluctant to view such transfers as a legitimate demand of justice.

Again we find a slippery-slope problem. Once we start down the road of equalizing natural endowments, where do we stop? Dworkin recognizes this slippery slope, and says that we might decide to draw an inviolable line around the body, regardless of how little any particular part of it is important to us, in order to ensure that the principle of equalizing circumstances does not violate our person. Libertarians, in practice, simply extend this strategy. If we can draw a line around the person, in order to ensure respect for individual personality, why not draw a line around her circumstances as well? In order to ensure that we do not end up with identical personalities due to identical upbringing, why not say that differential circumstances do not give rise to enforceable moral claims?

If we view libertarianism in this way, its popularity becomes more understandable. It is inhumane to deny that unequal circumstances create unfairness, and the attempts by libertarians to show that poverty is not a restriction on freedom or self-ownership just reveal how weak their defence of the free market is. But until we can find a clear and acceptable line between choices and circumstances, there will be some discomfort at making these forms of unfairness the basis of enforceable claims. Libertarianism capitalizes on that discomfort, by suggesting that we can avoid having to draw that line.

NOTES

1. It is particularly important to distinguish libertarians from 'neo-conservatives', even though both were part of the movement for free-market policies under Thatcher and Reagan, and so are sometimes lumped together under the label the 'New Right'. As we will see, libertarianism defends its commitment to the market by appeal to a broader notion of personal freedom—the right of each individual to decide freely how to employ her powers and possessions as she sees fit. Libertarians therefore support the liberalization of laws concerning homosexuality, divorce, abortion, etc., and see this as continuous with their defence of the market. Neo-conservatives, on the other hand, 'are mainly interested in restoring traditional values . . . strengthening patriotic and family feelings, pursuing a strong nationalist or anti-Communist foreign policy and reinforcing respect for authority', all of which may involve limiting 'disapproved lifestyles' (Brittan 1988: 213). The neo-conservative endorses market forces 'more because of the disciplines they impose than the freedom they provide. He or she may regard the welfare state, permissive morality, and "inadequate" military spending, or preparedness to fight, as different examples of the excessive self-indulgence that is supposed to be sapping the West.' From the libertarian point of view, therefore, neo-conservatives are the 'New Spartans', and the chauvinistic foreign policy and moralistic social policy adopted by Reagan and Thatcher stand opposite to their commitment to personal freedom (Brittan 1988: 240–2; cf. Carey 1984).

2. It is unclear whether Nozick himself would accept the claim that treating people as 'ends in themselves' is equivalent to treating them 'as equals', or whether he would accept Dworkin's egalitarian plateau. Rawls ties the idea of treating people as ends in themselves to a principle of equality (Rawls 1971: 251–7), and Kai Nielsen argues that

Dworkin's egalitarian plateau 'is as much a part of Nozick's moral repertoire' as Rawls's (Nielsen 1985: 307). However, even if there is some distance between Nozick's 'Kantian principle' of treating people as ends in themselves and Dworkin's principle of treating people as equals, they are clearly related notions, and nothing in my subsequent arguments requires any tighter connection. All that matters, for my purposes, is that Nozick defends libertarianism by reference to some principle of respect for the moral status and intrinsic worth of each person.

3. Nozick's claim is ambiguous here. He does not tell us what the 'normal process' of appropriation is. Hence it is unclear whether 'not worsening' is merely a necessary condition for legitimate appropriation (in addition to the 'normal process'), or whether it is a sufficient condition (any process which does not worsen the conditions of others is legitimate). If it is not a sufficient condition, he does not tell us what is (Cohen 1986a: 123).

4. Earlier libertarians recognized the insurmountable difficulties in justifying unequal appropriation of the initially unowned world, and many of them (reluctantly) accepted nationalization of the land (Steiner 1981: 561–2; Vogel 1988). Even Locke seemed to think that unequal property ownership could not arise from any right of individual appropriation. It required collective consent, in the form of an acceptance of money (Christman 1986: 163). In his survey of contemporary libertarianism, Norman Barry argues that none of the different versions of libertarianism (utilitarian, contractarian, natural rights, egoistic) have an adequate account of original title (Barry 1986: 90–3, 100–1, 127–8, 158, 178).

5. Andrew Kernohan argues that self-ownership does tell us something about property-ownership. He argues that some of the rights entailed in self-ownership logically entail access to resources. Owning one's powers, in the fullest legal sense, entails owning the exercise of these powers, and this requires the right to exercise those powers oneself, the managerial right to decide who else may exercise them, and the income right to any benefit which flows from their exercise. None of these rights can be fulfilled without some rights over resources (Kernohan 1988: 66–7). However, this logical connection between self-ownership and property-ownership still leaves a wide range of legitimate property-regimes. Indeed, the only regime it excludes is precisely the one Nozick wishes to defend—i.e. one where some people lack any access to resources. According to Kernohan, this lack of property-ownership is a denial of their self-ownership.

6. For futile attempts to show that mutual advantage is compatible with, and indeed requires, compulsory aid to the defenceless, see Lomasky (1987: 161–2, 204–8); Waldron (1986: 481–2); Narveson (1988:

269–74); Grice (1967: 149). For a discussion of their futility, see Goodin (1988: 163); Copp (1990); Gauthier (1986: 286–7).

7. Even if we can identify such norms of justice, there remains the difficult question of why we feel obliged to obey them. Why should I care about what I morally ought to do? Mutual advantage theorists argue that I only have a reason to do something if the action satisfies some desire of mine. If moral actions do not increase my desire-satisfaction, I have no reason to perform them. This theory of rationality may be true even if there are objective moral norms. Rawls's contractarianism may give a true account of justice, and yet 'be only an intellectual activity, a way of looking at . . . the world that can have no motivational effect on human action' (Hampton 1986: 32). Why should people who possess unequal power refrain from using it in their own interests? Buchanan argues that the powerful will treat others as moral equals only if they are 'artificially' made to do so 'through general adherence to internal ethical norms' (Buchanan 1975: 175–6). And indeed Rawls does invoke 'adherence to internal ethical norms'—namely a pre-existing disposition to act justly—in explaining the rationality of moral action (Rawls 1971: 487–9). In calling this 'artificial', Buchanan implies that Rawls has failed to find a 'real' motivation for acting justly. But why should not our motivation for acting justly be a moral motivation? As Kant put it, morality 'is a sufficient and original source of determination within us'. People can be motivated to act justly simply by coming to understand the moral reasons for doing so. This may seem 'artificial' to those who accept a mutual advantage view of rationality, but the acceptability of that view is precisely what is at issue. For many of us the recognition that others are fundamentally like ourselves in having needs and goals gives us a compelling reason to adopt the point of view of justice (Barry 1989: 174–5, 285–8).

8. Left-wing theorists often make the same mistake. George Brenkert, for example, argues that Marx's commitment to freedom is not tied to any principle of equality (Brenkert 1983: 124, 158; but cf. Arneson 1981: 220–1; Geras 1989: 247–51).

9. As these examples show, our interest in the freedom to do x is not simply our interest in doing x. I may care about the freedom to choose my own clothes, for example, even though I do not particularly care about choosing clothes. While my wardrobe is a matter of almost complete indifference, I would find any attempt by others to dictate my clothing to be an intolerable invasion of privacy. On the other hand, I may care about other freedoms, like the freedom to buy foreign goods without tariffs, only in so far they enable me to buy more goods. In yet other cases, our being free to do something, like religious worship, may be constitutive of the very value of that act. That we

freely choose to celebrate religious belief is crucial to the value of religious celebration. So our interest in the freedom to do x may be instrumental to, intrinsic to, or quite independent of, our interest in x. Hence our interest in different freedoms varies, not only with our interest in each particular act, but also with the range of instrumental, intrinsic, and symbolic interests promoted by having the freedom to do that particular act. Needless to say, it is hopelessly confusing to say that all these different interests are really a single interest in a more extensive freedom.

10. Nozick's claim here is not actually true. His theory does require that people's free exchanges preserve a particular pattern—namely, the Lockean proviso—and so it too requires continuously intervening in free exchanges to preserve a patterned distribution. This undermines Nozick's famous contrast between 'patterned theories', like Rawls's, and 'historical theories', like his own. All theories include both patterned and historical elements. Rawls, for example, allows people to come to have legitimate entitlements in virtue of their past actions and choices in conformity with the difference principle (a historical element), and Nozick requires that the pattern of distribution resulting from people's actions make no one worse off than they would have been in the state of nature (a patterned element). Nozick claims that the Lockean proviso is not a patterned requirement (Nozick 1974: 181), but if so, then nor is Rawls's difference principle (Bogart 1985: 828–32; Steiner 1977: 45–6). In any event, even if this contrast can be sustained, it is not a contrast between theories which interfere in people's lives and those which do not.

11. Gordon's subsequent discussion manifests these dangers. For example, he says that the free market increases people's freedom, but must be constrained in the name of justice. But he does not specify which people acquire which freedoms in the free market (specifying these things, he says, would obscure the problem of 'freedom as such'). As a result, he ignores the loss of freedom caused by private property, and hence creates a false conflict between justice and freedom. For a similarly confused attempt to preserve the idea of 'freedom' as a separate value, see Raphael (1970: 140–1). He notes that a redistribution of property could be seen as redistributing freedom in the name of justice, rather than as sacrificing freedom for justice. But, he says, this would eliminate freedom as a separate value, and so 'it is more sensible to acknowledge the complexity of moral objectives to be pursued by the State, and to say that justice and the common good are not identical with freedom, although they are all closely related', and hence 'the State ought *not* to intervene in social life to the utmost extent in order to serve the objectives of justice and the common good' (Raphael 1970: 140–1). In order to preserve the alleged contrast

between freedom and justice or equality, both Gordon and Raphael distort or ignore the actual freedoms and unfreedoms involved. Other discussions of what it might mean for liberty to be 'accorded priority over other political goods or values' rest on similar confusions—e.g. invoking criteria to measure freedom that appeal to these other values, thus rendering the priority claim unintelligible (e.g. Gray 1989: 140–60; Loevinsohn 1977).

5

Marxism

The standard left-wing critique of liberal justice is that it endorses formal equality, in the form of equal opportunity or equal civil and political rights, while ignoring material inequalities, in the form of unequal access to resources. This is a valid criticism of libertarianism, given its commitment to formal rights of self-ownership at the expense of substantive self-determination. But contemporary liberal egalitarian theories, like those of Rawls and Dworkin, do not seem vulnerable to the same criticism. Rawls does believe that material inequalities (under the difference principle) are compatible with equal rights (under the liberty principle), and some critics take this as evidence of a lingering commitment to formal equality (e.g. Daniels 1975: 279; Nielsen 1978: 231; Macpherson 1973: 87–94). But the inequalities licensed by the difference principle are intended to promote the material circumstances of the less favoured. Far from neglecting substantive self-determination in the name of formal equality, the difference principle is justified precisely because 'the capacity of the less fortunate members of society to achieve their aims would be even less' were they to reject inequalities which satisfy the difference principle (Rawls 1971: 204). To oppose these inequalities in the name of people's substantive self-determination is therefore quite misleading.[1]

Given this shared commitment to material equality, do socialists and liberal egalitarians share the same account of justice? For some strands of socialist thought, the answer is yes. There seems to be no deep difference between Dworkin's liberal theory of equality of resources and various socialist theories of 'compensatory justice', which also aim at an ambition-sensitive, endowment-insensitive distribution (e.g. Dick 1975; DiQuattro 1983; cf. Carens 1985). But there are other strands of socialist thought which move in a different direction. I will discuss two such strands in this chapter, both of which are found in recent Marxist writings. One strand objects to the very idea of justice. Justice, on this view, is a remedial virtue, a response to some flaw in social life. Justice seeks to mediate conflicts between individuals, whereas communism overcomes those conflicts, and hence overcomes the need for justice. The second strand shares liberalism's emphasis on justice, but rejects the liberal belief that justice is compatible with private ownership of the means of production. Within this second strand, there is a division between those who criticize private property on the grounds of exploitation, and those who criticize it on the grounds of alienation. In either case, however, Marxist justice requires socializing the means of production, so that productive assets are the property of the community as a whole, or of the workers within each firm. Where liberal egalitarian theories of justice try to employ private property while negating its inequalities, Marxists appeal to a more radical theory of justice that views private property as inherently unjust.

1. COMMUNISM AS BEYOND JUSTICE

One of the most striking features of Rawls's theory is its claim that 'justice is the first virtue of social institutions' (Rawls 1971: 3). Justice, according to Rawls, is not one amongst a number of other political values, like freedom, community, and efficiency. Rather, justice is the standard by which we weigh the importance of these other values. If a policy is unjust, there is no separate set of values one can appeal to in the hope of overriding justice, for the legitimate weight attached to these other values is established by their location within the best theory of justice. (Conversely, one of the tests of a theory of justice is that it gives due weight to these other values. As Rawls says, if a theory of justice does not give adequate scope to community and freedom, then it will not be attractive to us.)

Liberals emphasize justice because they see a tight connection between it and the basic idea of moral equality. Liberals promote the moral equality of people by formulating a theory of juridical equality, which articulates each individual's claims to the conditions which promote their well-being. Many Marxists, on the other hand, do not emphasize justice, and indeed object to the idea that communism is based on a principle of justice. In this regard, they are following Marx himself, who attacked the ideas of 'equal right' and 'fair distribution' as 'obsolete verbal rubbish' (Marx and Engels 1968: 321). This is the conclusion Marx draws from his analysis of the 'contribution principle'—i.e. the claim that labourers have a right to the products of their labour. While many socialists in his day viewed the contribution principle as an important argument for socialism, Marx says that it has many 'defects' which make it, at best, a transitional principle between capitalism and communism. The contribution principle gives people an 'equal right', since everyone is measured by an equal standard (i.e. labour). However, some people have greater natural talents, so this equal right becomes an 'unequal right for unequal labour':

> it tacitly recognizes unequal individual endowment and thus productive activity as natural privileges. *It is, therefore, a right of inequality, in its content, like every right.* Right by its very nature can consist only in the application of an equal standard; but unequal individuals (and they would not be different individuals if they were not unequal) are measurable only by an equal standard in so far as they are brought under an equal point of view, are taken from one *definite* side only, for instance, in the present case, are regarded *only as workers* and nothing more is seen in them, everything else being ignored. (Marx and Engels 1968: 320)

According to Allen Wood, this passage shows that Marx was averse, not only to the idea of justice, but to the concept of moral equality underlying it. On Wood's view, Marx was 'no friend to the idea that "equality" is something good in itself', and he was not 'a believer in a society of equals' (Wood 1979: 281; 1981: 195; cf. Miller 1984: ch. 1).

But Marx's argument here does not reject the view that the community should treat its members as equals. What he denies is that the community should do so through implementing a theory of juridical equality. In this passage, Marx endorses a principle of equal regard, but denies that any 'equal right' ever captures it because rights work by defining one limited viewpoint from which individuals are to be regarded equally. For example, the contribution principle views people as workers only, but ignores the fact that different workers vary both in their talents and in their needs— for example, 'one worker is married, another not; one has more children than another, and so on' (Marx and Engels 1968: 320). In reality, the number of viewpoints relevant to determining the true meaning of equal regard is indefinite or, in any event, cannot be specified in advance. But notice that this description of the effect of 'equal rights' is only a criticism if people are owed equal concern and respect—that is why these inequalities are 'defects'. Marx rejected the idea of equal rights, not because he was not a friend to the idea of treating people as equals, but precisely because he thought rights failed to live up to that ideal. In fact, the idea of moral equality is basic to Marx's thought (Arneson 1981: 214–16; Reiman 1981: 320–2; 1983: 158; Geras 1989: 231, 258–61; Elster 1983: 296; 1985: ch. 4).[2]

Marxists have a number of objections to the idea of juridical equality. The first one, we have seen, is that equal rights have unequal effects, since they only specify a limited number of the morally relevant standpoints. But that argument is weak, for even if it is true that we cannot define in advance all the relevant standpoints, it does not follow that the best way of treating people with equal regard is by not specifying any viewpoints at all. Even if a schedule of rights cannot fully model equal regard, it may do so better than any other alternative. In fact, what else can we do except try to specify the standpoints we think morally relevant? We can only avoid this difficult task by avoiding having to make distributive decisions at all. This is indeed what some Marxists have hoped to do, by assuming that there will be an abundance of resources under communism, but, as we will see, this is an unrealistic hope.

A second objection is that theories of 'just distribution' concentrate too much on *distribution*, rather than on the more fundamental questions of *production* (Young 1981; Wood 1972: 268; Buchanan 1982: 56–7, 122–6; Wolff 1977: 199–208; Holmstron 1977: 361; cf. Marx and Engels 1968: 321). If all we do is redistribute income from those who own productive assets to those who do not, then we will still have classes, exploitation, and hence the kind of contradictory interests that make justice necessary in the first place. We should instead be concerned with transferring ownership of the means of production themselves. When this is accomplished, questions of fair distribution become obsolete.

This is an important point. Our concern should be with ownership, for ownership allows people not only to accrue greater income, but also to gain a measure of control over other people's lives. A scheme of redistributive taxation may leave a capitalist and a worker with equal incomes, but it would still leave the capitalist with the power to decide how the worker spends much of her time, a power that the worker lacks in relation to the capitalist. As an objection to the idea of justice, however, this complaint fails. Nothing in the idea of justice limits it to questions of income. On the contrary, as we have seen, both Rawls and Dworkin include productive assets as one of society's resources to be distributed in accordance with a theory of justice. Indeed, Rawls argues that a more egalitarian pattern of property-ownership is required for his ideal of a 'property-owning democracy'. And if Dworkin tends, when discussing the practical implementation of his theory, to look at schemes of income redistribution, rather than a fundamental redistribution of wealth, that is incompatible with his theory of justice (ch. 3, s. 5 above). The Marxist objection to the class structure of capitalist relations of production is, above all, a distributive objection, and so fits comfortably within the normal scope of theories of justice (as Marx himself sometimes noted— Marx and Engels 1968: 321; Marx 1973: 832; cf. Arneson 1981: 222–5; Geras 1989: 228–9; Cohen 1988: 299–300).

At best, these two objections point to limitations in the way that some people have developed their conceptions of justice. The heart of the Marxist critique, however, is an objection to the very idea of a juridical community. Marxists believe that justice, far from being the first virtue of social institutions, is something that the truly good community has no need for. Justice is appropriate only if we are in the 'circumstances of justice', circumstances which create the kinds of conflicts that can only be solved by principles of justice. These circumstances are usually said to be of two main kinds: conflicting goals, and limited material resources. If people disagree over goals, and are faced with scarce resources, then they will inevitably make conflicting claims. If, however, we could eliminate either the conflicts between people's goals, or the scarcity of resources, then we would have no need for a theory of juridical equality, and would be better off without it (Buchanan 1982: 57; Lukes 1985: ch. 3).

According to some Marxists, the circumstance of justice which communism seeks to overcome is conflicting conceptions of the good. They take the (idealized) family as an example of an institution which is non-juridical, where there is an identity of interests, in which people respond spontaneously to the needs of others out of love, rather than responding on the basis of rightful duties or calculations of personal advantage (cf. Buchanan 1982:

13). If the community as a whole also had an identity of interests and affective ties, then justice would not be needed, because to conceive of oneself as a bearer of rights is to 'view oneself as a potential party to interpersonal conflicts in which it is *necessary* to assert claims and to "stand up" for what one claims as one's due' (Buchanan 1982: 76). If we fulfilled each other's needs out of love, or out of a harmony of interests, then there would be no occasion for such a concept of rights to appear.

I have argued elsewhere that Marx did not believe in this vision of an affectively integrated community with an identity of interests. For Marx, communist relations are free of antagonism 'not in the sense of individual antagonism, but of one arising from the social conditions of life of the individuals' (Marx and Engels 1968: 182).[3] The 'harmony of ends' solution to the circumstances of justice is, in fact, more of a communitarian ideal than a Marxian one (cf. ch. 6, s. 4c below). Moreover, it is doubtful that this is a possible solution to the circumstances of justice. For even if we share a set of goals, we may still have conflicting personal interests (e.g. two music-lovers wanting the only available opera ticket). And even where we lack conflicting personal interests, we may disagree about how to achieve a shared project, or about how much support it deserves. You and I may both believe that experiencing music is a valuable part of a good life, and that music should be supported with one's time and money. But you may wish to support music in such a way as to allow the greatest number of people to experience it, even if that means that they experience lower-quality music, whereas I want to support the highest-quality music, even if that means some people never experience it. So long as there are scarce resources, we will disagree over how much support should go to which musical projects. Shared ends only eliminate conflicts over the use of scarce resources when people share means and priorities as well. But the only people who share identical ends for the identical reasons with identical intensity are identical people. And this raises the question of whether conflicting ends are best seen as a 'problem' which needs to be 'remedied' or overcome. It is perhaps true that conflicts are not, in and of themselves, something to be valued. But the diversity of ends which makes such conflicts inevitable may be something to be valued.

The other solution to the circumstances of justice is to eliminate material scarcity. As Marx puts it:

In a higher phase of communist society, after the enslaving subordination of the individual to the division of labour . . . has vanished; after labour has become not only a means of life but life's prime want; after the productive forces have also increased with the all-round development of the individual, and all the springs of co-operative wealth flow more abundantly—only then can the narrow horizon of bourgeois right be crossed in its entirety and society inscribe on its banners: From each according to his ability, to each according to his needs! (Marx and Engels 1968: 320–1)

Marx was emphatic about the need for abundance, for he thought that scarcity made conflicts unresolvable. The highest development of the productive forces 'is an absolutely necessary practical premise [of communism] because without it *want* is merely made general, and with *destitution* the struggle for necessities and all the old filthy business would necessarily be reproduced' (Marx and Engels 1970: 56). It was perhaps because he was so pessimistic about the social effects of scarcity that he became so optimistic about the possibility of abundance (Cohen 1990*b*).

However, this solution to the circumstances of justice is also implausible (Lukes 1985: 63–6; Buchanan 1982: 165–9; Nove 1983: 15–20). Certain resources (e.g. space) are inherently limited, and the recent wave of environmental crises has revealed the empirical limits to other resources we depend on (e.g. depleting oil reserves). Moreover, certain kinds of conflicts and harms can arise even with an abundance of certain resources. One example that arises because of an ability and desire to help others is the potential conflicts involved in paternalism. So even if justice is appropriate only as a response to problems in society, it may not be possible to overcome these problems.

But is justice best viewed as a remedial virtue that should be superseded? Marxists argue that while justice helps mediate conflicts, it also tends to create conflicts, or at any rate, to decrease the natural expression of sociability. Hence justice is a regrettable necessity at present, but a barrier to a higher form of community under conditions of abundance. It is better if people act spontaneously out of love for each other, rather than viewing themselves and others as bearers of just entitlements.

But why are these two opposed, why must we choose either love or justice? After all, some people argue that a sense of justice is a precondition of, and indeed partly constitutive of, love for others.

The Marxist assumption seems to be that if we give people rights they will automatically claim them, regardless of the effects on others, including the ones they love. For example, Buchanan says that justice involves 'casting the parties to conflict in the narrow and unyielding roles of rights-bearers' (Buchanan 1982: 178; cf. Sandel 1982: 30–3). But why can I not choose to waive my rights whenever their exercise would harm the people I love? Consider the family. Does the fact that women in France now have the right to move to another town and work there without their spouse's permission mean that they will exercise that right rather than keep their families together? (Similarly, have men, who have always had that right, never forgone a career move for the sake of their families?) Buchanan says that 'for those who find the bonds of mutual respect among right-bearers too rigid and cold to capture some of what is best in human relationships, Marx's vision of genuine community—*rather than* a mere juridical association—will remain attractive' (Buchanan 1982: 178, my emphasis). But if the family is an example of what is best in human relationships, then the contrast is spurious. The family has always been a juridical association, in which spouses and children are all rights-bearers (though not equally so). Does that mean that marriage is after all not a sphere of mutual affection, but, as Kant put it, an agreement between two people for 'reciprocal use of each other's sexual organs'? Of course not. Families can have loving relationships, and the juridical nature of marriage does nothing to prevent them. Surely no one believes that people will only act out of love if they are denied the opportunity to do otherwise.

Rawls's claim about the priority of justice is not a claim 'about whether a person will, or should, push to the limit their rightful claims to various advantages' (Baker 1985: 918). While the priority of justice ensures that individuals are able to claim certain advantages, it equally ensures that they are able to share these advantages with those they love. Generous and loving people will be generous and loving with their just entitlements—far from inhibiting this, the priority of justice makes it possible. What justice excludes is not love or affection, but injustice—the subordination of some people's good to others', through the denial of their just entitlements (Baker 1985: 920). And this, of course, is the opposite of genuine love and affection.

Justice is not only compatible with a concern for others, it is itself an important form of concern for others. It is often said that a concern for rights involves a self-understanding that is grounded in egoism and a concern to protect oneself against the likely antagonism of others in a zero-sum social world. Buchanan, for example, says that to think of oneself as a rights-bearer is to 'view oneself as a potential party to interpersonal conflicts in which it is *necessary* to assert claims' (Buchanan 1982: 76). To claim a right, on this view, is to have a certain pessimistic view of how others will respond to our requests. But Buchanan himself suggests another reason for valuing the recognition of rights. He says that for someone to think of himself as a rights-bearer is to 'think of himself as being able to demand what he has a right to as his due, rather than as something he may merely request as being desirable' (Buchanan 1982: 75–6). These are two very different self-understandings, despite the conflation by Buchanan. The second self-understanding concerns not the probability of getting something I want or need, but the grounds on which I think of myself as properly (e.g. not selfishly) having it. I may want to avoid taking advantage of the (potentially self-sacrificing) love of others. If so, then justice can serve as a standard for determining what I am non-selfishly entitled to, even when others are prepared to give me more than I am entitled to.

Justice can also serve as a standard for determining how to respond to the needs of others, even where the reason I want to help is simply my love for them. I may desire to help several people, all of whom are in need, because I love them, not because I owe them my help. But what if their needs conflict? As Rawls points out, it is no good to say I should act benevolently rather than justly, for 'benevolence is at sea as long as its many loves are in opposition in the persons of its many objects' (Rawls 1971: 190). While love is my motivation, justice may be the standard I appeal to, given that love yields conflicting imperatives. Hence, 'while friendship may render justice unnecessary as a *motive*, it may still require some aspects of justice as a *standard*. Friends do not automatically know what to do for one another' (Galston 1980: 289 n. 11). Justice, therefore, serves two important purposes. When making claims, I may wish to know what I am properly entitled to, even when others will fulfil my requests without concern for my entitlements. When responding to the claims of others, I may wish to know what their entitlements are, even when love is the motivation for my response.

In neither case is my interest in the teachings of justice a matter of 'standing up for my due'.

The public recognition of rights can be valuable in another way. A person may be sure of getting what she wants, by virtue of her participation in a highly respected social practice (e.g. as a teacher). She has no need for rights, on Buchanan's first account of them, because others value her contribution to that practice, and reward her generously for it. Yet she may want to know that people would accord her rights even if they did not share her commitments. And she may want to know this even if she has no desire to leave that practice, for this recognizes that she is a source of value in and of herself, not just *qua* occupant of a social role.

Justice is more than a remedial virtue. Justice does remedy defects in social co-ordination, and these defects are ineradicable, but it also expresses the respect individuals are owed as ends in themselves, not as means to someone's good, or even to the common good. Justice recognizes the equal standing of the members of the community, through an account of the rights and entitlements we can justly claim. But it does not force people to exercise these entitlements at the expense of the people or projects they care about. Justice constitutes a form of concern that we should have for the members of our community, and enables us to pursue all the other forms of love and affection which are consistent with that underlying moral equality. The view that we could create a community of equals by abandoning these notions of fairness, rights, and duties is untenable.[4]

2. COMMUNIST JUSTICE

If justice is both ineradicable and desirable, what would Marxist justice look like? It is standardly supposed that Marxism is egalitarian, indeed more egalitarian than liberalism, further to the left. This is certainly true in regard to mainstream liberalism and its ideology of equal opportunity, according to which unlimited inequalities are legitimate so long as there is fair competition for higher-paying positions (ch. 3, s. 2 above). But it is not immediately obvious what room there is to the left of Rawls's version of liberal egalitarian justice, since it too rejects the prevailing ideology, and it places no limits on the extent to which resources should be equalized. What distinguishes Marxist from Rawlsian justice is not the extent to which resources should be equalized, but rather the form in which such equalization should occur. Rawls believes that equality of resources should take the form of equalizing the amount of private property available to each person. For Marx, on the other hand, 'the theory of the Communists may be summed up in a single phrase: Abolition of private property'. Private ownership is permissible in areas of 'personal property', like the clothes, furnishings, and leisure goods we use at home and at play. But it is 'fundamental' to Marxism that 'there is no moral right to the private ownership and control of productive resources' (Geras 1989: 255; cf. Cohen 1988: 298). Equalization of productive resources should take the form of socializing the means of production, so that each person has equal participation in collective decisions about the deployment of productive assets, made at the level of either individual firms or national economic planning.

Why should equality take the form of equal access to public resources, rather than an equal distribution of private resources? One reason is simply that Rawls's idea of a 'property-owning democracy' may not be empirically viable. It may be impossible to equalize productive resources in modern economies except through socializing ownership. As Engels put it, 'the bourgeoisie . . . could not transform these puny means of production into mighty productive forces without transforming them, at the same time, from means of production of the individual into *social* means of production only workable by a collectivity of men'. Under capital-ism, however, these 'socialized means of production' are still treated 'just as they had been before, i.e., as the means of production of individuals'. The solution to this contradiction 'can only come about by society openly and directly taking possession of the productive forces which have outgrown all control except that of society as a whole' (Marx and Engels 1968: 413, 414, 423).

For Engels, the need to socialize ownership is not based on any distinctive theory of justice, but simply on an inability to conceive of any other device for equalizing resources in a modern industrial economy. Some Marxists also object on empirical grounds to Rawls's assumption that the inequalities arising from market transactions in a well-ordered society would tend to benefit the less well off. If they would not (and Rawls gives no evidence that they would), and if redistributive mechanisms are inherently vulnerable to political pressure, then we might adopt socialism on the basis of a 'greater-likelihood principle' (Schweickart 1978: 11, 23; DiQuattro 1983: 68–9; Clark and Gintis 1978: 324).

These claims suggest that Rawls's idea of a property-owning democracy is 'at best fanciful' (Nielsen 1978: 228). Indeed, some critics argue that the whole idea of a property-owning democracy only makes sense in its original Jeffersonian context of an agrarian society composed of independent landholders (Macpherson 1973: 135–6; Weale 1983: 57). If so, then socializing the means of production may be the only viable way of implementing the difference principle. Unfortunately, Rawls has not sufficiently developed his model of a just society for us to evaluate its viability.

While these objections to the viability of an egalitarian private property regime account for much of the left-wing criticism of Rawls's theory—and for much of the day-to-day debate between liberal egalitarians and socialists—there are also more theoretical objections to the very idea of private property. According to many Marxists, private ownership of the means of production should be abolished because it gives rise to the wage-labour relationship, which is inherently unjust. Some Marxists claim that wage-labour is inherently exploitative; others claim that it is inherently alienating. On either view, justice is only secured by abolishing private property, even if a Rawlsian property-owning democracy is empirically viable.

(a) Exploitation

The paradigm of injustice for Marxists is exploitation, and, in our society, the exploitation of the worker by the capitalist. The fundamental flaw of liberal justice, Marxists claim, is that it licenses the continuation of this exploitation, since it licenses the buying and selling of labour. Does liberal justice allow some to exploit others? It depends, of course, on how we define exploitation. In its everyday usage, exploitation (when applied to persons rather than natural resources) means 'taking unfair advantage of someone'. Every theory of justice, therefore, has its own theory of exploitation, since every theory has an account of the ways it is permissible and impermissible to benefit from others. On Rawls's theory, for example, a talented person takes unfair advantage of the untalented if he uses their weak bargaining position to command an unequal share of resources not justified by the difference principle. It is not exploitative, however, for someone to benefit from employing others if this works to the maximal benefit of the least well off. If we are convinced of the fairness of Rawls's theory, then we will deny that it licenses exploitation, since part of what is involved in accepting a theory of justice is accepting its standard for judging when others are unfairly taken advantage of.

Marxists, however, operate with a more technical definition of exploitation. In this technical usage, exploitation refers to the specific phenomenon of the capitalist extracting more value from the worker's labour (in the form of produced goods) than is paid back to the worker in return for that labour (in the form of wages). According to classical Marxist theory, capitalists only hire workers when they can extract this 'surplus value', and so this exploitative transfer of surplus value from the worker to the capitalist is found in all wage-relationships. This technical definition of exploitation is sometimes said to be of scientific rather than moral interest. For example, the fact that capitalists extract surplus value is said to explain how profits are possible in a competitive economy, and this claim does not by itself entail that it is wrong to extract surplus value. Most Marxists, however, have taken the extraction of surplus value as evidence of an injustice—indeed, as the paradigm of injustice.

Does Marxist exploitation have moral significance, i.e. does it involve taking unfair advantage of someone? The traditional argument that technical exploitation is unjust goes like this (from Cohen 1988: 214):

1. Labour and labour alone creates value.
2. The capitalist receives some of the value of the product.

Therefore:

3. The labourer receives less value than he creates.
4. The capitalist receives some of the value the labourer creates.

Therefore:

5. The labourer is exploited by the capitalist.

There are a number of gaps in this argument. Premiss (1) is controversial, to say the least. Many Marxists have tried to defend it by appeal to 'the labour theory of value', according to which the value of a produced object is determined by the amount of labour required to produce it. But as Cohen points out, the labour theory of value actually contradicts (1), for the labour theory says that the value of an object is determined by the amount of labour currently required to produce it, not how much labour was actually involved in producing it. If technology changes in such a way that an object can now be produced with half the labour previously required, the labour theory of value says that the value of the object is cut in half, even though the amount of labour embodied in the already produced object is unaffected. The actual labour expended by the worker is irrelevant, if the labour theory of value is true.

What matters, morally speaking, is not that the workers create value, but that 'they create *what has value*. . . . What raises a charge of exploitation is not that the capitalist appropriates some of the value the worker produces, but that he appropriates some of the value of *what* the worker produces' (Cohen 1988: 226–7). Creating products that have value is different from creating the value of those products, and it is the former that really matters for the charge of exploitation. Even if someone other than the worker creates the value of the product—if, for example, its value is determined by the desires of its consumers—then Marxists would still say that the worker is exploited by the capitalist, for it is the worker, not the capitalist or the consumers, who created the product. Hence the proper argument is this (Cohen 1988: 228):

1'. The labourer is the only person who creates the product, that which has value.
2. The capitalist receives some of the value of the product.

Therefore:

3'. The labourer receives less value than the value of what he creates.
4'. The capitalist receives some of the value of what the labourer creates.

Therefore:

5. The labourer is exploited by the capitalist.

This modified version of the Marxist argument yields the conclusion that wage-relationships are inherently exploitative. But it is not clear that the exploitation involved here is an injustice. In the first place, there is nothing unjust about volunteering to contribute one's labour to others. Most Marxists, therefore, add the proviso that the worker must be forced to work for the capitalist. Since workers do not in general own any productive assets, and can only earn a living by working for a propertied capitalist (though not necessarily for any particular capitalist), most wage-relationships fall under this proviso (Reiman 1987: 3; Holmstrom 1977: 358).

Is the forced transfer of surplus value exploitative in the everyday sense? This is both too weak and too strong. It is too weak in excluding from the purview of exploitation wage-labour which is not, strictly speaking, forced. If, for example, a safety net is in place, guaranteeing a minimal income to all, then the propertyless can acquire a subsistence living through the welfare state, without having to work for a capitalist. But we might still want to say that workers are exploited. While the propertyless are not forced to work for a capitalist in order to survive, that may be the only way for them to earn a decent standard of living, and we might think it is unfair that they should have to yield surplus labour to capitalists in order to secure a comfortable living. One can say that such people are 'forced' to work for the capitalist, since the alternatives are in some way unacceptable or unreasonable. But, as we will see, the important question is not whether workers are forced to work for capitalists, but whether the unequal access to resources which 'forces' workers to accept that surplus transfer is unfair.

Defining exploitation as forced transfer of surplus labour is also too strong, for there are many legitimate instances of forced transfer of surplus value. What if workers are like apprentices who must work for others for a period of five years, but then are able to become capitalists themselves (or masters)? According to Jeffrey Reiman, this is exploitative: 'We care about workers being forced to sell their labor power, because we understand this as forcing them to work without pay. And we care about how long workers are forced to work without pay, because of how we feel about people being forced to work without pay for any period of time' (Reiman 1987: 36). But this is implausible. If all workers can become capitalists, and if all capitalists begin as workers, then there is no inequality over the course of people's lifetimes. Like apprentices, there is simply a period where workers have to pay their dues (Cohen 1988: 261 n. 9). To insist that it is exploitative to transfer surplus value forcibly, regardless of how this fits into a larger pattern of distributive justice, guts the charge of exploitation of all its moral force. It manifests a kind of fetishism about owning one's labour. Indeed, it manifests a libertarian concern with self-ownership:

Marxists say that capitalists steal labour time from working people. But you can steal from someone only that which properly belongs to him. The Marxist critique of capitalist injustice therefore implies that the worker is the proper owner of his own labour time: he, no one else, has the right to decide what will be done with it. . . . Hence the Marxist contention that the capitalist exploits the worker depends on the proposition that people are the rightful owners of their own powers. . . . [Indeed], if, as Marxists do, you take appropriation of labour time as such, that is, in its fully general form, as a paradigm of injustice, then you *cannot* eschew affirmation of something like the self-ownership principle. (Cohen 1990*a*: 366, 369)

That this is a libertarian assumption is shown by the fact that compulsory taxation to support children or the infirm also counts as exploitation, according to Reiman's definition. If we force workers to pay taxes to support the infirm, then we are forcing them to work without pay.[5]

In his initial presentation of the Marxist exploitation argument, Cohen denied that it presupposes that people own the products of their labour: 'One can hold that the capitalist exploits the worker by appropriating part of the value of what the worker produces without holding that all of that value should go to the worker. One can affirm a principle of distribution according to need, and add that the capitalist exploits the worker because need is not the basis on which he receives part of the value of what the worker produces' (Cohen 1988: 230 n. 37). But what then is the justification for saying that the capitalist is exploiting the worker? Assuming that the capitalist does not need the object, and hence has no legitimate claim to it, it does not follow that the worker has any claim under the needs principle. The person most in need may be some third party (e.g. a child), and then the child has the only legitimate claim to the object. If the capitalist none the less appropriates the object, then he is unjustly treating the child, not the worker. Indeed, if the worker appropriates the object, then she too is unjustly treating the child. When the needs principle is violated, the people who are unjustly treated are the needy, not the producers.

Moreover, what if the capitalist does need the surplus value? Let us say that the capitalist is infirm, and has had the good fortune to have inherited a large number of shares in a company. Cohen implies that this is still exploitation, for 'his need is not the basis on which he receives part of the value of what the worker produces'. Rather, it is his ownership of the means of production. But the worker's need is not the basis on which she receives the product either. Rather, it is her production of the object. So who then is the capitalist exploiting? No one, for under the needs principle, no one else had any legitimate claim to the resources. Moreover, why cannot need be the basis on which a capitalist receives surplus value? What if the government, in order to avoid leaving support for the infirm subject to the vagaries of day-to-day politics, endows the infirm with capital from which they can derive a steady stream of financial support? Distributing capital to the infirm might, in fact, be a very good way of meeting the needs principle (cf. Cohen 1990*a*: 369–71; Arneson 1981: 206–8). Once we drop the self-ownership claim, then the appropriation of surplus labour is not, as such, inherently exploitative—it all depends on how the particular transaction fits into a larger pattern of distributive justice.

There is another problem with the exploitation argument. What about those who are forced not to sell their labour? Married women have been legally precluded from taking wage employment in many countries. Hence they are not exploited. On the contrary, they are being protected from exploitation, which is indeed how many people defend sexual discrimination. But if married women in these countries are given a small income from government taxes, then they become exploiters, on the Marxian exploitation argument, since part of each worker's income is forcibly taken away and put at their disposal. But it would be perverse to view women under these circumstances as beneficiaries of exploitation. They suffer from an injustice worse than exploitation by capitalists, and one of the first tasks of feminist movements has been to gain equal access for women to wage-labour markets. Or consider the unemployed, who are legally able to accept wage employment, but can find none. They too are not exploited, under the Marxist definition, since they do not produce any surplus value for the capitalist to appropriate. And if the government taxes workers to pay them a benefit, then they too become exploiters. Yet they are worse off than those who are able to find a wage relationship (Roemer 1982*b*: 297; 1988: 134–5).

These examples show that there is a deeper injustice underlying exploitation—namely, unequal access to the means of production.

Disenfranchised women, the unemployed, and wage-workers all suffer from this injustice, while capitalists benefit from it. The exploitation of workers by capitalists is just one form this distributive inequality can take. The subordinate positions of women and the unemployed are other forms, and judging by people's struggles to gain wage employment, these may be more damaging forms. For those who lack access to property, being forced to sell one's labour may be better than being forced not to (women), or being unable to (unemployed), or eking out a marginal existence from crime, begging, or living off whatever land remains common property (Marx's 'lumpenproletariat').

Something has gone wrong here. Exploitation theory was supposed to provide a radical critique of capitalism. Yet, in its standard form, it neglects many of those who are worst off under capitalism, and actually precludes the action needed to help them (e.g. welfare support for children, the unemployed, and the infirm). If exploitation theory is to take due account of these groups, it must abandon the narrow focus on surplus transfer, and instead examine the broader pattern of distribution in which these transfers occur. This is the main aim of John Roemer's work on exploitation. He defines Marxist exploitation, not in terms of surplus transfer, but in terms of unequal access to the means of production. Whether one is exploited or not, on his view, depends on whether one would be better off in a hypothetical situation of distributive equality— namely, where one withdrew with one's labour and per capita share of external resources. If we view the different groups in the economy as players in a game whose rules are defined by existing property-relations, then a group is exploited if its members would do better if they stopped playing the game, and withdrew with their per capita share of external resources and started playing their own game. According to Roemer, both employed and unemployed workers would be better off by withdrawing from the capitalist game, and so are exploited.

Exploitation in the technical sense—the transfer of surplus value—plays only a limited role in Roemer's theory. It is one of the most common results of distributive injustice under capitalism, but it has no ethical interest apart from that inequality. It is 'a bad thing only when it is the consequence of an *unjust* unequal distribution in the means of production' (Roemer 1988: 130). Surplus transfer is legitimate when it is untainted by distributive inequality, or when it helps compensate for that inequality. For example, state-mandated support for the unemployed and for disenfranchised women reduces, rather than creates, exploitation, for it helps rectify 'the loss suffered by [them] as a result of the unequal initial distribution of property' (Roemer 1988: 134). For Roemer, the 'ethical imperative' of exploitation theory, therefore, is not to eliminate surplus transfers, but to 'abolish differential ownership of the alienable means of production' (Roemer 1982*b*: 305; 1982*c*: 280).

Cohen says that Roemer's theory makes Marxists 'more consistent egalitarians' (Cohen 1990*a*: 382). But Roemer's account of Marxist exploitation still views compulsory support of the infirm (or children) as exploitation, for it gives them more than they would be able to secure for themselves with their per capita share of resources.[6] Inequalities due to unequal natural talents are not a matter of Marxist exploitation, and so Roemer's 'ethical imperative' is still less egalitarian than those theories which attempt to compensate for natural disadvantages. By defining exploitation in terms of the results of an unequal distribution of external resources, Roemer works 'without recourse to the radical egalitarian premiss of denying self-ownership' (Roemer 1988: 168). He expresses sympathy with theories which take this radical step, like those of Rawls and Dworkin. And he himself takes this step when he says that differences in natural talents are a source of 'socialist exploitation' (i.e. a form of exploitation that continues to exist under socialism). But while he opposes socialist exploitation, and endorses restrictions on self-ownership, he says this is a separate issue from specifically 'Marxist exploitation', which presupposes that people are entitled to the fruits of their labour (Roemer 1982*c*: 282–3; 1982*b*: 301–2). Marxist exploitation theory works with the 'more conservative' premiss that people have rights of self-ownership, so that equality of resources does not include any requirement that unequal talents be compensated for (Roemer 1988: 160; cf. 1982*a*: chs. 7–8).

Arneson gives a similar account of exploitation. Like Roemer, he says that judgements of 'wrongful exploitation' require a comparison with a hypothetical egalitarian distribution. And like Roemer's account of socialist exploitation (as opposed to specifically Marxist exploitation), Arneson's account of equal distribution precludes differences that arise from unequal natural talents as well as unequal external resources. Arneson believes that most workers under capitalism are exploited according to this test, for they suffer from undeserved inequalities in either wealth or talent which enable others to take advantage of them (Arneson 1981: 208). As with Roemer, surplus transfer plays a derivative role in Arneson's theory. Surplus transfer is wrong if it is the result of an unequal distribution, but is legitimate if it arises independently of, or if it is used to compensate for, undeserved differences in wealth or natural talents. Hence compulsory support for the unemployed is legitimate (unlike Reiman), as is support for the infirm (unlike Roemer's 'Marxist exploitation'). Most of the surplus taken from workers under capitalism, however, is not of the legitimate kind, since it winds up in the hands of those who benefit from the unequal distribution of talents and wealth. Hence capitalism is exploitative, albeit for more complex reasons than suggested by the initial Marxian exploitation argument.

This is a more plausible account of exploitation. By focusing on the broader pattern of distribution, not just the exchange that occurs within the wage-relationship, Roemer and Arneson avoid both of the problems that plagued Reiman's account. Their accounts allow us to say that workers in a welfare state can be exploited, whether or not they are forced to work for capitalists, since they are denied fair access to the means of production. Their accounts also allow us to deal with cases of distributive injustice that occur outside the wage-relationship, like the injustice of being unable or unfree to find employment, since these too involve a denial of fair access to resources.

Unfortunately, this is a more attractive approach precisely because it has left behind all that was distinctive about the original Marxist approach to exploitation. This new approach differs in three important ways from that original approach. Firstly, the idea of exploitation is now derived from a prior and broader principle of distributive inequality. In order to know what counts as exploit-ation, we need first to know what people are entitled to by way of rights over themselves and external resources. And once we make these underlying principles explicit, it is clear that exploitation is simply one of many forms of distributive injustice, not the paradigm of injustice. Unfortunately, Marxists remain prone to exaggerating the moral centrality of exploitation. Roemer, for example, expands the ambit of exploitation to cover all forms of distributive inequality.[7] As we have seen, this enables him to consider the fate of the unemployed as well as the wage-worker. But it is confusing to call both of these cases of exploitation. Our everyday sense tells us that exploitation requires some direct interaction between exploiter and exploited in which the former takes unfair advantage of the latter, and this is not generally true of the unemployed. They are unfairly neglected or excluded, but not necessarily unfairly taken advantage of, for capitalists may gain no benefit from their plight. To say that all forms of injustice are forms of exploitation is not to gain an insight but to lose a word. Moreover, Roemer's attempted assimilation obscures the relationship between equality and exploitation. He says that different forms of inequality (unfair advantage, exclusion, neglect) are all cases of the broader category of exploitation. But the opposite is more accurate—exploitation is one of the many forms of inequality, all of which are assessed by a deeper and broader principle of equality. On Roemer's theory, this deeper principle of equality is expressed in the 'ethical imperative' to equalize access to resources. Exploitation is no longer at the moral heart of the theory.

Secondly, the broader theory of justice in which exploitation is situated has become progressively closer to a Rawlsian theory of justice. The original Marxist argument said that workers are entitled to the product of their labour, and it is the forced denial of that entitlement which renders capitalism unjust. But most contemporary Marxists have tried to avoid that libertarian premiss, since (among other reasons) it makes aid to the dependent morally suspect. And the more they try to accommodate our everyday sense that not all technical exploitation is unjust, the more they have appealed to Rawlsian principles of equality. While the Marxist rhetoric of exploitation is taken to be more radical than liberal egalitarian views of justice, 'the Marxist condemnation of the injustice of capitalism is not so different from the conclusion that other apparently less radical and contemporary theories of political philosophy reach, albeit in language less flamboyant than Marxism's' (Roemer 1988: 5). Arneson's theory of Marxist exploitation, for example, appeals to the same principle of an ambition-sensitive, endowment-insensitive distribution that underlies Dworkin's theory. In its new forms, Marxist exploitation theory seems to apply liberal egalitarian principles, rather than compete with them.

Finally, this new account of exploitation abandons what was the *raison d'être* of the original Marxist exploitation argument—

namely, the claim that there is an inherent injustice in wage-labour. For if the test of wrongful exploitation is whether there are undeserved inequalities, then some wage-relationships are not exploitative. There are two 'clean routes' to wage-labour. Firstly, as we have seen, endowing the infirm with ownership of capital can compensate for unequal natural talents, and so bring us closer to an endowment-insensitive distribution. Secondly, differential owner-ship of the means of production can arise amongst people with equal endowments, if they have different preferences concerning investment or risk. In the gardener–tennis-player example I used in chapter 3, the tennis-player wanted to use his resources immedi-ately in consumption, in the form of a tennis-court, whereas the gardener invested her resources in production, in the form of a vegetable garden (ch. 3, s. 3*b*(ii) above). This was legitimate, I argued, even though the tennis-player ends up working for the gardener (or some other owner of productive assets), because it met the 'envy test'. Each party was free to make the same choices as the other, but neither party desired the other's lifestyle, since they had different preferences about work and leisure. Similarly, the gardener might have acquired more assets by taking a large risk, while the tennis-player, who could have taken the same risk, preferred to have a smaller but risk-free income. Different choices about leisure and risk can lead, in a legitimate and envy-free way, to unequal ownership of productive assets. Where people's preferences do not differ in these ways, or where any such differences are less important to people than a shared desire to have a democratic say in one's workplace, then we are likely to maintain a system of equal ownership of productive assets. But to enforce a blanket pro-hibition on wage-labour would be an arbitrary violation of the ambition-sensitivity requirement of a just distribution.

None of this justifies existing inequalities in ownership of the means of production. Marx scorned those who argued that capitalists acquired their property through conscientious savings, and he went on to show that 'conquest, enslavement, robbery, murder, in short, force, play the greatest part' in capital accumul-ation (Marx 1977*c*: 873–6, 926; cf. Roemer 1988: 58–9). This unjust initial acquisition undermines the risk argument, for even if capitalists are willing to take risks with their capital, it is not (morally speaking) their capital to take risks with. Workers might be willing to take the same risks as capitalists if they had any capital to take risks with. In any event, 'it cannot seriously be maintained that a worker's life involves less risk than a capitalist's. Workers face the risk of occupational disease, unemployment, and impoverished retirement, which capitalists and managers do not face' (Roemer 1988: 66). So neither effort nor risk-adversity can justify existing inequalities (Roemer 1982*b*: 308; *contra* Nozick 1974: 254–5). But the fact that capitalism historically arose out of undeserved inequalities does not show that wage-labour could not arise legitimately within a regime such as Rawls's 'property-owning democracy'. Indeed, if people are well informed about the consequences of their choices, and if their different preferences were formed under conditions of justice, then 'the argument appears irrefutable' (Elster 1983: 294).[8]

So private property need not be exploitative. Conversely, socialization of the means of production may be exploitative. Marxists are fond of saying that exploitation is impossible within socialism, since producers control their product (e.g. Holmstrom 1977: 353). But on the new approach to exploitation, it is not enough that people have equal access to social resources, in the form of a vote in a democratically run, worker-owned firm. It all depends on what people democratically decide to do with their resources. Consider a firm that is permanently divided into two groups—a majority which, like the gardener, prefers income to leisure, and a minority which, like the tennis-player, prefers leisure to income. If the majority wins all the decisions, and if the minority are not allowed to convert their socialist right of equal access to social resources into a liberal entitlement of equal individual resources (e.g. by selling their share of the firm), they will be unfairly taken advantage of. They will be exploited, on the Roemer–Arneson approach, since they would be better off by withdrawing with their per capita share of resources (Arneson 1981: 226; Geras 1989: 257).

Concern for exploitation, therefore, does not justify a general preference in favour of socializing, rather than equalizing, the means of production. Equalizing resources may be non-exploitative, even if some people work for others, and socializing resources may be exploitative, even if everyone works for each other. It depends on the preferences people have, and the circumstances they find themselves in. What matters is that people have the sort of access to resources that enables them to make whatever decisions concerning work, leisure, and risk best suit their goals in life. That kind of self-determination may be best achieved through a mix of private property, public ownership, and worker democracy, since each form of ownership creates certain options while blocking others (Lindblom 1977: ch. 24; Goodin 1982: 91–2; Weale 1982: 61–2). These are largely technical questions, and they cannot be pre-empted by blanket charges of exploitation.

(b) Needs

So far, I have not said much about Marx's claim that distribution under communism will be based on the principle 'to each according to his needs'. I did say that this principle is incompatible with the traditional Marxist conception of exploitation, which excludes the compulsory transfer of surplus labour from workers to others. But what can be said of it on its own, as a principle of justice? As we have seen, it is possible that Marx himself did not think of it as a principle of justice. Given his prediction of abundance, 'to each according to his needs' is not a principle under which scarce resources are distributed, but simply a description of what happens under communism—people take what they need from the stock of abundant resources (Wood 1979: 291–2; Cohen 1990*b*; but cf. Geras 1989: 263).

Most contemporary Marxists, however, do not share Marx's optimism concerning abundance, and instead invoke the needs principle as a distributive principle. Viewed in this way, it is most plausibly understood as a principle of equal need-satisfaction, since Marx offers it as a solution to the 'defects' of the contribution principle, which, we have seen, are the inequalities created by people's different needs (Elster 1983: 296; 1985: 231–2). Is this an attractive principle? It is not very attractive if needs are interpreted in terms of bare material necessities. A socialist government that only provided for people's bare material needs would hardly constitute an advance on the welfare-state programmes of some Western democracies. Marxists, however, interpret 'needs' in a much more expansive way. Indeed for Marx, human needs are distinguished by their 'limitless and flexible nature', so that people's needs include 'a rich individuality which is as all-sided in its production as in its consumption' (Marx 1977*c*: 1068; 1973: 325). Hence 'needs' is being used as a synonym for 'interests', which include both material necessities and the various goods that people feel worth having in their lives. So construed, needs encompass such things as important desires and ambitions, and so the needs principle 'is most plausibly understood as a principle of equal welfare' (Elster 1983: 296), rather than a principle of equal need-satisfaction in any more limited sense.

Unfortunately, once we adopt this expansive interpretation of the needs principle, it no longer gives us much guidance on how to distribute resources. Marxists seem to think that the needs principle is an answer to the question of what it means to give equal consideration to people's interests. But once we expand 'needs' to encompass all our interests, and drop the assumption of abundance, saying that distribution ought to be by need is not an answer to that question, but simply another way of asking it. It tells us nothing about how to attend to the different sorts of interests we have. For example, while needs in the minimal sense are not matters of choice, needs in the Marxist sense fall on both sides of the choices–circumstances line. Whether a given share of resources satisfies one's needs, therefore, depends on how expensive one's needs are, which depends on both one's circumstances and one's choices. Should we provide extra resources for those with expensive needs? And if so, should we spend all our resources on the severely handicapped? Should we distinguish between expensive needs that are chosen and those that are not? These are the questions that Rawls and Dworkin focus on, for without an answer to them, a theory of justice is incomplete. But Marxists have not, in general, explained how the needs principle assigns weight to people's interests.

In so far as Marxists have given some content to the needs principle, the most common area of disagreement with liberal egalitarianism concerns the claim that people are responsible for the costs of their choices, and hence that distributions should be ambition-sensitive. Some Marxists reject the claim outright, on the grounds that people's choices are the product of their material or cultural circumstances, so that people are not responsible for their choices (e.g. Roemer 1985a: 178–9; 1986: 107, 109; 1988: 62–3). Levine says that Roemer's denial that individuals are responsible for their choices 'suggests a far more radical conception of what it is to treat people as equals' than is found in Dworkin's theory (Levine 1989: 51 n. 25). But it is not clear what is particularly radical (or attractive) about requiring some people to subsidize other people's expensive tastes, and many Marxists would see this as unfair. As Arneson puts it, 'Consider two persons, both with artistic need, one of whom is cost-conscious and learns to satisfy this need through media that are cheap (watercolors, pen-and-ink drawings), while the other is not mindful of cost and develops talents that can be exercised only at extravagant cost (huge marble sculptures, deep-sea photography). It is not obvious that "to each according to his need" is the appropriate principle for distributing scarce social resources to these artists' (Arneson 1981: 215). In order to deal with expensive choices, the needs principle requires some guidelines regarding what counts as 'reasonable' needs, so that 'people could be told at the early stages of preference formation that society will not underwrite all sorts of expensive tastes' (Elster 1983: 298; Geras 1989: 264). According to Arneson, the need for such a social norm reflects 'the vagueness of Marx's slogan' but does not 'call in question its basic moral thrust' (Arneson 1981: 215). But as Elster notes, this is in fact a very different understanding of the needs principle, since it tells people to adjust their needs to a pre-existing standard of distribution, whereas the needs principle is usually interpreted as requiring that we adjust distribution to people's pre-existing needs (Elster 1983: 298).

Whether this ambition-sensitivity requirement is best seen as an abridgement of the needs principle or as an elaboration of it, the net effect is to make Marxist equality more or less identical to Dworkin's theory of equality of resources (Elster 1983: 298 n. 65).[9] Or if it is different, Marxists do not tell us how it differs, for they do not tell us how to measure the costs of people's choices. What, for example, plays the role of Dworkin's auction? Marxists have traditionally opposed market mechanisms. But if people are to be held responsible for the costs of their choices, then something like a market is required to measure that opportunity cost. (See Nove 1983: part 1, on how the Marxist hostility to markets, combined with the assumption of abundance, has prevented Marxists from coming up with any coherent notion of opportunity costs.)

There is less dispute over the claim that a just distribution should be endowment-insensitive. The needs principle 'severs all connections between the amount of benefits one receives from the economy and the "morally arbitrary" genetic and social factors

that determine one's ability to contribute to that economy'
(Arneson 1981: 215–16). The requirements of the needs principle
are clearer here, for it was 'designed precisely to take care of such
instances' (Elster 1983: 298). But even here the needs principle is
incomplete, for it does not tell us what to do when it is impossible
to compensate fully for natural disadvantages. As we saw in
chapter 3, it is impossible to equalize the circumstances of a
multiply handicapped person, and undesirable to devote all our
resources to that task. This led Dworkin to devise his hypothetical
insurance scheme. But there is no similar solution to this problem
in the contemporary Marxist literature, and indeed no similar
recognition that it is a problem. It is not enough to say that the
needs principle compensates for unequal circumstances. We need to
know how to do so, and at what price. Given these unanswered
questions, the needs principle does not compete with liberal
theories of equality, for it is simply at a less-developed stage of
formulation.

(c) Alienation

If Marxists are committed to abolishing private property, they must
appeal to something other than exploitation. According to Steven
Lukes, Marx's critique of capitalism appeals not only to a 'Kantian'
concern with exploitation, but also to a 'perfectionist' concern with
alienation (Lukes 1985: 87; cf. Miller 1989: 52–4).[10] Whereas the
Kantian strand emphasizes the way private property reduces some
people (the workers) to a means for the benefit of others
(capitalists), the perfectionist strand emphasizes the way private
property inhibits the development of our most important capacities. The problem with private property is not simply that it is
exploitative, for even those who benefit from exploitation are
alienated from their essential human powers. This alienation
argument seems a more promising route for defending a prohibition on private property, for while equalizing private property
eliminates exploitation, it may just universalize the alienation.

Perfectionist arguments, of which Marx's alienation argument is
one example, say that resources should be distributed in such a way
as to encourage the 'realization of distinctively human potentialities
and excellences', and to discourage ways of life which lack these
excellences (Lukes 1985: 87). Such theories are 'perfectionist'

because they claim that certain ways of life constitute human 'perfection' (or 'excellence'), and that such ways of life should be promoted, while less worthy ways of life should be penalized. This is unlike liberal or libertarian theories, which do not try to encourage any particular way of life, but rather leave individuals free to use their resources in whatever ways they themselves find most valuable. I will consider the general contrast between liberal and perfectionist theories in the next chapter, but I will look briefly at how Marxist perfectionism might defend a prohibition on private property.

Any perfectionist argument must explain what the 'distinctive human excellences' are, and how the distribution of resources should be arranged so as to promote them. In Marx's case, our distinctive excellence is said to be our capacity for freely creative co-operative production. To produce in a way that stunts this capacity is to be 'alienated' from our true 'species-nature'. Hence, Marxist perfectionists argue, resources in a communist society should be distributed so as to encourage people to achieve self-realization through co-operative production. Distribution might still be governed by the needs principle, but for perfectionists the needs principle is not concerned with all needs. Rather, it would involve 'some selection of those forms of human interest and concerns which most fully express the ideal of co-operative, creative, and productive activities and enjoyments' (Campbell 1983: 138; cf. Elster 1985: 522).

How should this ideal be promoted? Marxists argue it is best promoted by abolishing wage-labour and socializing the means of production. Wage-labour alienates us from our most important capacity, because it turns the worker's labour-power into a mere commodity the disposition of which is under someone else's control. Moreover, for many workers under capitalism, this exercise of labour-power tends to be mindless and devoid of any intrinsic satisfaction. Socializing the means of production ensures that each person has an effective say in how her work life is organized, and enables her to organize production so as to increase its intrinsic satisfaction, rather than to increase the profits of the capitalist. Capitalism reduces our life's activity to a means which we endure in order to secure a decent living, but socialism will restore work to its rightful place as an end in itself, as 'life's prime want' (or, more accurately, socialism will make it possible for the first time in history for labour to assume this rightful place).

This, then, is the perfectionist argument for abolishing private property in the means of production. What are we to make of it? Phrased as a choice between intrinsically satisfying and intrinsically unsatisfying work, most people will favour creative and co-operative work. The evidence is overwhelming that most workers in capitalism wish that their jobs were more satisfying. The 'degradation of labour' which capitalism has imposed on so many people is abhorrent, an unconscionable restriction on their ability to develop their human potential (Schwartz 1982: 636–8; Doppelt 1981). Liberals try to deal with this by distinguishing legitimate and illegitimate ways that people can come to be employed by others. But on the Marxist view, any wage-relationship is alienating, since the worker gives up control over her labour-power, and over the products of her labour. Wage-labour may not be exploitative, if both parties started with an equal share of resources, but it is alienating, and we can eliminate the alienation by socializing productive resources rather than equalizing private property.

However, while unalienated labour is surely better than alienated labour, these are not the only values involved. I may value unalienated labour, yet value other things even more, such as my leisure. I may prefer playing tennis to unalienated production. I must engage in some productive work to secure the resources necessary for my tennis, and all else being equal, I would prefer it to be unalienated work. But all else is not always equal. The most efficient way to produce goods may leave little room for creativity or co-operation (e.g. assembly-line production). If so, then engaging in non-alienated work may require a greater investment in time than I am willing to make. For example, if I can acquire the resources I need by doing either two hours a day of alienated work or four hours a day of unalienated work, the extra two hours of tennis may outweigh the two hours of alienation. The question, then, is not whether I prefer unalienated labour to alienated labour, but whether I prefer leisure so much that I would accept alienated labour in order to acquire it. Opportunities for unalienated work 'are not so much manna from the sky. Resources must be used to make these opportunities available, which means lesser availability of some other goods', like leisure (Arneson 1987: 544 n. 38).[11]

Consumption is another good that may conflict with non-alienated production. Some people enjoy consuming a wide variety of goods and services, from food to opera to computers. Agreeing to perform alienated labour in return for higher wages may enable them to expand their range of desired consumption. If we prohibit alienated labour, we eliminate their alienation, but we also make it more difficult for them to pursue forms of consumption they truly value. Marxist perfectionists tend not to be concerned with possible decreases in material consumption. They consider people's concern with consumption as a symptom of the pathology of materialism created by capitalism, so that the transition to socialism 'will involve a large shift in cultural emphasis from consumption to production as the primary sphere of human fulfillment' (Arneson 1987: 525, 528). But is it pathological to be concerned with expanding one's consumption? The 'keeping up with the Joneses' syndrome may be, for the pursuit of such status goods is often irrational. But that is not true of many desires for increased consumption. There is nothing pathological about a music-lover wanting expensive stereo equipment, and being willing to perform alienated labour to acquire it. Hence there is no reason for communism to 'exclude or stigmatize those who prefer the passive pleasures of consumption' over the active pleasures of production (Elster 1985: 522).

The pursuit of unalienated labour can also conflict with relationships with family and friends. I may want a part-time job that allows me as much time as possible with my children, or perhaps seasonal work, so that I can spend part of each year with friends or relatives. As Elster notes, the Marxist emphasis on self-realization in work can compete with spontaneous personal relationships, for there is a 'tendency for self-realization to expand into all available time . . . [and this] is a threat both to consumption and to friendship' (Elster 1986: 101).

The issue is not whether unalienated labour is a good, but whether it is an overriding good, a good which is necessary to any decent life, and which outweighs in value all competing goods. There is no reason whatsoever to think unalienated labour is such a good. Marx's own argument for this claim reveals how implausible it is. He argued that freely co-operative production is our distinctive human excellence because this is what differentiates us from other species—it is what defines us as humans. But this 'differentia' argument is a *non sequitur*. Asking what is best in a human life is not a question 'about biological classification. It is a question in moral philosophy. And we do not help ourselves at all in answering it if we decide in advance that the answer ought to be a single, simple characteristic, unshared by other species, such as the differentia is meant to be' (Midgley 1978: 204). Exaltation of co-operative productive activity 'is a particular moral position and must be defended as such against others; it cannot ride into acceptance on the back of a crude method of taxonomy' (Midgley 1978: 204). Whether or not other animals have the same capacity for productive labour as humans has no bearing on the question of the value of that capacity in our lives. There is no reason to think that our most important capacities are those that are most different from other animals.

This focus on productive labour is also sexist. Consider Marx's claim that because workers are alienated from their 'species-life' (i.e. 'labour, life activity, productive life itself'), therefore 'man (the worker) only feels himself freely active in his animal functions— eating, drinking, procreating, or at most in his dwelling and in dressing-up, etc.; and in his human functions he no longer feels himself to be anything but an animal' (Marx 1977*a*: 66). But why is production a more 'human function' than reproduction (e.g. raising children)? It may be less distinctively human, in the sense that other animals also reproduce. But this just shows how irrelevant that criterion is, for family life is surely as important to our humanity as production. Marx combined a profound sensitivity to historical variations in the predominantly male sphere of productive life with an almost total insensitivity to historical variations in the pre-dominantly female sphere of reproductive life, which he viewed as essentially natural, not distinctively human (Jaggar 1983: ch. 4; O'Brien 1981: 168–84). Any theory which hopes to incorporate the experience of women will have to question the elevation of productive labour.

There are many values that may compete with unalienated production, such as 'bodily and mental health, the development of cognitive facilities, of certain character traits and emotional responses, play, sex, friendship, love, art, religion' (Brown 1986: 126; cf. Cohen 1988: 137–46). Some people will view productive labour as 'life's prime want', but others will not. A prohibition on alienated labour, therefore, would unfairly privilege some people over others. As Arneson puts it, the identification of socialism with a particular vision of the good life 'elevates one particular category of good, intrinsic job satisfaction, and arbitrarily privileges that good and those people who favour it over other equally desirable goods and equally wise fans of those other goods' (Arneson 1987: 525). Given that people differ in the value they attach to labour, 'differential alienation of labour, from an initial position of equal opportunity and fair division of assets, can vastly increase the welfare and life quality of people'. Hence 'a perfectionist defence of nonalienation seems remote' (Roemer 1985*b*: 52).

Not all Marxists who emphasize the flourishing of unalienated production under communism are perfectionists. Some Marxists who proclaim the end of alienation are simply making a prediction about what people will do with their equal resources, not giving a perfectionist instruction about how to distribute those resources. They predict that people will value unalienated labour so highly that they will never accept improved leisure or family life as compensation for alienation. Should this prediction turn out to be false, however, there would be no reason to interfere with people's choices by prohibiting alienation. It is unclear whether Marx's comments on alienation are predictions or perfectionist instructions (Arneson 1987: 521). Engels, however, was anti-perfectionist, at least in the case of sexual relations. When discussing the nature of sexual relations in communism, he says that the old patriarchal relations will end, but

what will there be new? That will be answered when a new generation has grown up: a generation of men who never in their lives have known what it is to buy a woman's surrender with money or any other social instrument of power; a generation of women who have never known what it is to give themselves to a man from any other considerations than real love or to refuse to give themselves to their lover from fear of the economic consequences. When these people are in the world, they will care precious little what anybody today thinks they ought to do; they will make their own practice and their corresponding public opinion about the practice of each individual—and that will be the end of it. (Engels 1972: 145)

The equal distribution of resources ensures that exploitative relations will not arise, but there is no correct socialist model of personal relations which is to be encouraged or imposed. But why should not economic relations likewise be left to the free choices of people from a position of material equality? We should wait and see what that 'new generation' will choose to do with their lives and talents, and while they may systematically favour unalienated labour, there is no reason for perfectionist intervention to encourage that result.

Again, none of this will justify the existing distribution of meaningful work. I have argued that people should be free to sacrifice the quality of work life for other values, like better leisure. Under capitalism, however, those with the best jobs also have the best consumption and leisure, while those with poor jobs get no compensating increase in leisure or consumption. But the solution is not to give everyone the best possible work, at the expense of improved leisure, since some people would rather have better leisure. As Arneson puts it, 'The core socialist objection to a capitalist market is that people who have fewer resources than others through no fault of their own do not have a fair chance to satisfy their preferences. The solution to this problem is not to privilege anybody's preferences [e.g. those for work over leisure], but to tinker with the distribution of resources that individuals bring to market trading.' Hence Dworkin's aim of an envy-free market distribution 'is one aspect of socialist aspiration, not a rival doctrine' (Arneson 1987: 537, 533).

This returns us to the 'Kantian' strand of Marxist thought, which leaves individuals free to decide for themselves what is worth doing with their fair share of resources. And, we have seen, this leads to a series of questions about fair distribution which Marxists simply have not addressed. Until they do, it is difficult to tell whether Marxism provides a distinctive account of justice from those of other political traditions.

3. THE POLITICS OF MARXISM

Both strands of Marxist thought share a preoccupation with labour. The Kantian strand views work as the fundamental site of capitalist injustice (i.e. exploitation). The perfectionist strand views work as the fundamental site of the socialist goal of non-alienation. But there is a third sense in which labour is fundamental to Marxism—namely, the fact that workers are identified as the main agents of social change. According to Marxist sociology, the struggle against capitalist injustice will take the form of a struggle between two increasingly polarized classes—workers and capitalists. Capitalists must oppress workers, for their wealth comes from the exploitation of workers, and workers must oppose capitalists, since they have nothing to lose but their chains. Class conflict is endemic to the wage-relationship, which is endemic to capitalism, and so the wage-relationship is the linchpin around which revolutionary struggle occurs. Other groups may be unfairly treated, but Marxists have viewed them as marginal in terms of both power and motivation. Only workers are able and willing to challenge the whole edifice of capitalist injustice. To concentrate on the fate of other groups is reformist, not revolutionary, since their oppression is less, and less essential, than the workers'.

Marxist theories of justice are, in large part, attempts to give the rationale for this class-struggle. As Roemer puts it, 'the purpose of a theory of exploitation is . . . to explain class struggle. As Marxists, we look at history and see poor workers fighting rich capitalists. To explain this, or to justify it, or to direct it and provide it with ideological ammunition, we construct a theory of exploitation in which the two antagonistic sides become classified as the exploiters and the exploited' (Roemer 1982c: 274–5). And since the explanation of class-struggle is located directly in the wage-relationship, there has been a natural tendency for Marxists to locate the justification for socialism directly in the wage-relationship. Thus we get theories of the inherent exploitation or alienation of wage-labour.

It is increasingly difficult to accept this traditional Marxist view about the centrality of labour to progressive politics. Many of the most important contemporary struggles for justice involve groups which are not, or not only, oppressed by the wage-relationship— e.g. racial groups, single mothers, immigrants, gays and lesbians, the disabled, the elderly. As we have seen, support for these groups may in fact conflict with the labour-emphasizing arguments for socialism. Marxists have tended in practice to support the claims of these non-proletarian groups. They have tended to support the needy, whether or not their needs are related to any labour-based principle of alienation or exploitation. But as Cohen points out, they have often justified doing so by treating 'the set of exploited producers as roughly coterminous with the set of those who needed the welfare state's benefits' (Cohen 1990a: 374). In other words, while Marxist theory has been labour-based, its practice has been needs-based, and the obvious inconsistencies have been papered over by assuming that the needy are also the exploited.

However, it is increasingly clear that the needy and the Marxian exploited are not always the same people. This 'forces a choice between a principle of self-ownership embedded in the doctrine of exploitation and a principle of equality of benefits and burdens that negates the self-ownership principle and which is required to defend support for very needy people who are not producers and who are, a fortiori, not exploited' (Cohen 1990a: 13–14). I have argued that it is arbitrary, at the level of theory, to endorse the 'fetishism of labour' implicit in the doctrines of exploitation and alienation (Roemer 1985b: 64). But it is also unhelpful at the level of practice, for it neglects forms of injustice which motivate some of the most important contemporary progressive political movements. If there is to be an effective movement for radical social change, it will have to involve a coalition of both the needy and the exploited. But the rhetoric of Marxist exploitation and alienation does not speak to the needs of non-labourers, and may indeed oppose them.

Marxists pride themselves on their unity of theory and practice. But their theory betrays their practice. Faced with the choice between self-ownership and distributive equality, Marxists have, in practice, embraced equality, and have done so in a much more committed way than liberals have. But at the level of theory, Marxists remain committed to a fetishism of labour that is in some ways less radical, and less attractive, than liberal egalitarian theories of justice, and this has hampered the quest for an effective radical movement. A genuine unity of theory and practice may require a greater unity of Marxism and liberal equality.

NOTES

1. While both liberal egalitarians and Marxists share a commitment to material equality, they disagree over the means which can be used to pursue it. If a society is in violation of the difference principle, but respects civil rights, then Rawls and Dworkin deny that we can limit civil liberties in order to correct the material inequality. As I mentioned in chapter 3, they advance radical goals, but reject radical means to achieve them (see ch. 3 n. 9).
2. The idea of moral equality often appears in Marx in the same form in which it appears in Kant, Rawls, and Nozick—i.e. as the requirement that we should treat people as ends in themselves, not means. He

thought that capitalism failed to treat people as ends both in the relations of production (where the capitalist labour process reduces the worker to the status of a thing, an instrument, to be exploited by the capitalist), and in the relations of exchange (where 'each views the needs and desires of the other not as needs and desires, but rather as levers to be manipulated, as weaknesses to be preyed upon'—Buchanan 1982: 39).

3. Marx says that life under communism would be a 'social life' (Marx 1977a: 90–3), and that communist individuals would be 'social individuals' (Marx 1973: 705, 832). But he did not say that there will be an inherent harmony of interests, nor that we should aim at creating such a harmony. For my view of what these claims about social life amount to, see Kymlicka (1989a: ch. 6).

4. Marxists have other objections to justice. For example, appeals to justice are said to be strategically divisive, due to the essential contestability of the idea of justice, and are unnecessary, since the motor of history is the rational interest of the disadvantaged. Also, conceptions of justice are ideological, shaped to suit existing property relations, and hence a socialist conception of justice must follow, rather than precede, changes in property-relations. Many of these supposed defects of justice are said to follow from Marx's theory of historical materialism. For the ideology objection, see Wood (1981: 131–2); Brenkert (1983: 154–5); Wood (1972: 274); and the response in Geras (1989: 226–8); Nielsen (1989); Arneson (1981: 217–22). For the role of moral motivation in class struggle, see Wood (1984); Miller (1984: 15–97); and the response in Geras (1989: 251–4); Nielsen (1987).

5. Reiman denies that compulsory aid to the disabled is exploitative because it can be seen as an insurance policy that everyone buys, and so is 'an indirect return to individuals of labor equal to what they contribute, and thus not altering the basic distributive principle' that no one is made to work for anyone else (Reiman 1989: 312 n. 12). But this is clearly false of many recipients of such aid—e.g. the congenitally infirm. Holmstrom, who like Reiman defines exploitation in terms of 'forced, unpaid, surplus labour' (Holmstrom 1977: 358), says that support for the infirm is not exploitative, because 'The surplus is under the control of those who produce it. There is no class of non-producers who appropriate what workers have produced. Workers do not consume it all, directly or indirectly, but they control it as a class' (Holmstrom 1977: 363). But the fact that workers control it as a class does not show that individual workers are not forced to hand over surplus product to others. What if I, as an individual worker, object to what the working class as a whole decides to do with the surplus product? Can I insist that I get the full value of what I have

produced? If not, and if I must work to earn a living, then I am exploited on her definition. Moreover, what if there is a constitutional guarantee of welfare rights, under which workers are legally required to support the infirm? Then, under her definition, the working class as a whole is exploited, since they do not legally control the entire surplus.

6. Roemer tries to avoid this implication by adding a 'dominance condition' (Roemer 1982*a*: 237), or the requirement that there be no 'consumption externalities' (i.e. the workers must not get any pleasure from helping the disabled) (Roemer 1989: 259). But these are *ad hoc*, since they are disconnected from the 'ethical imperative' he identifies as the basis of exploitation theory (as he admits—Roemer 1982*c*: 277 n.). Indeed, they seem to be question-begging attempts to disavow the libertarian content of exploitation theory, and to block libertarian claims that the welfare state is exploitative (Bertram 1988: 126–7).

7. Roemer has, in places, moved away from this assimilation of injustice and exploitation. In order to match our everyday sense that exploitation involves someone taking advantage of another, he adds the following proviso: not only must the exploited group fare better by withdrawing with its talents and per capita share of resources, the exploiters must fare worse if the exploited withdraw with their existing resources (Roemer 1982*b*: 285). Where this added condition is not met, groups that are denied equal access to resources are 'Marxian-unfairly treated', but not 'exploited', for others do not take advantage of them ('they could disappear from the scene and the income of the others would not change'—Roemer 1982*b*: 292). But, as he admits, this added condition still fails to capture our intuitive sense of 'taking unfair advantage' (Roemer 1982*b*: 304 n. 12; cf. Elster 1982: 366–9). In his recent book, Roemer returns to his original definition of exploitation as 'the loss suffered by a person as a result of the unequal initial distribution of property' (Roemer 1988: 134), whether or not this loss comes about from being taken advantage of. Hence a person is capitalistically exploited 'if he would gain by virtue of an egalitarian redistribution of society's alienable means of production' (Roemer 1988: 135), and the unemployed are as exploited as wage-workers under this test.

8. For Roemer's attempt to refute the argument, see Roemer (1988: 149–55). His main objection is that even if we equalized resources, the differential ownership of capital which would arise from people's choices would largely reflect the lingering influence of earlier injustice. Those people who were born into poor families would not be taught the habits of risk-taking and deferred gratification which are passed on within rich families. Different preferences concerning work and leisure do not justify differential ownership of the means of production, because the preferences themselves were formed under conditions of

injustice (Roemer 1988: 62–3, 152–3; 1985*b*: 52). This is a valid point—people are fully responsible for their choices only when their preferences are formed under conditions of justice (cf. Rawls 1979: 14; Arneson 1981: 205; Scanlon 1988: 185–201). But this hardly defends a blanket prohibition on private property. It suggests that, for a generation or two, we must attend to, and perhaps compensate for, this influence. Perhaps we could implement an affirmative action programme to encourage previously disadvantaged groups to acquire and pass on the relevant dispositions. This does not undermine the general principle that different ambitions can legitimately give rise to differential ownership of the means of production.

9. Some socialists who accept the principle that distribution should be ambition-sensitive are none the less concerned to limit the kinds of inequalities it generates. For example, some say that extensive differences in income would violate self-respect (Nielsen 1978: 230; Daniels 1975: 273–7; Doppelt 1981: 259–307; Keat 1982: 68–70; but cf. Rawls 1971: 107; DiQuattro 1983: 59–60; Gutmann 1980: 135–8), or would undermine the conditions necessary for developing a sense of justice (Clark and Gintis 1978: 315–16), or a sense of solidarity (Crocker 1977: 263). I doubt that these problems would arise for income differences which pass the envy test (how can the greater resources of others violate my self-respect when they ac-company a lifestyle that I did not want and freely rejected?). Some say that extensive income inequality would undermine the equality of political power necessary for democracy (Daniels 1975: 256–8), or would create unequal opportunities for children (Nielsen 1985: 297–8). These are serious worries, but they are recognized by Rawls and Dworkin, who agree that they impose constraints on legitimate inequalities (*re* political equality, see Rawls 1971: 225–6; Dworkin 1988; *re* unequal opportunity, see Rawls 1971: 73). For further socialist views on ambition-sensitivity, see Nielsen (1985: 293–302); Elster (1985: 231–2, 524); Levine (1988: 53).

Carens argues that the central difference between socialists and liberals involves, not the needs principle, but the other half of Marx's famous slogan ('from each according to his abilities, to each according to his needs'). Carens takes this as imposing a duty on people to contribute, a duty to make 'good use' of their talents, whereas liberals think this could enslave the talented by forcing them to do something they are good at but do not enjoy (Carens 1986: 41–5). I do not believe that most Marxists actually share Carens's duty-imposing interpretation of Marx's slogan, but it is an important issue that deserves more discussion.

10. Lukes also distinguishes a 'utilitarian' strand in Marx's thought, but I will leave this aside, partly because we have already examined utilitarianism, and partly because this strand of Marx's thought has

had less influence on contemporary Marxists than the Kantian and perfectionist strands. Moreover, I doubt there was a utilitarian strand in Marx's thought. He rejected the idea that a person can be harmed just because that would increase the overall good (Murphy 1973: 217–20; but cf. Allen 1973; Brenkert 1981).

11. Marx himself once claimed that 'the realm of freedom really begins only where labour determined by necessity and external expediency ends; it lies by its very nature beyond the sphere of material production proper' (Marx 1981: 958–9). This is not his usual view of the matter, nor is it shared by most contemporary Marxists (e.g. Cohen 1978: 323–5), but it is surely true that the 'development of human powers as an end in itself' can occur outside production, and that 'nothing in the nature of things prevents the sphere of leisure from becoming the main arena for that free many-sided self-development of the individual that Marx prized' (Arneson 1987: 526).

Even if we accept the emphasis on production as the arena of self-fulfilment, there are other values besides non-alienation at stake. Marxists say that the value of production lies in 'the development of human powers as an end in itself'. But some people think that the value of productive labour lies in contributing to an organization that efficiently serves vital needs. For such 'service-oriented' workers, workplace democracy may be a 'wasteful self-indulgence' which places the welfare of the worker above that of the recipients (Arneson 1987: 525). Perfectionists argue that work is only meaningful if there is worker democracy (Nielsen 1978: 239; Schwartz 1982: 644). But what is wrong with caring more about what gets done than about how it gets done? There is a plurality of goods to be gained from labour—Arneson lists 17 of them—of which the free development of one's talents is just one, and different goods flourish best under different systems of work-organization and property-ownership (Arneson 1987: 527). So there is no simple correlation between socializing productive assets and increasing the value of our productive activities.

6

Communitarianism

1. LIBERAL INDIVIDUALISM AND STATE NEUTRALITY

With the exception of the perfectionist strand of Marxism, all of the theories we have examined so far share an important assumption. While they disagree on how to show equal concern for people's interests, they agree on how to characterize those interests, or at least they agree on a central feature of that characterization. They all believe that we promote people's interests by letting them choose for themselves what sort of life they want to lead. They disagree about what package of rights and resources best enables people to pursue their own conceptions of the good. But they agree that to deny people this self-determination is to fail to treat them as equals.

I have not yet discussed the importance of self-determination (it is one of the issues I set aside in chapter 3 by postponing discussion of Rawls's liberty principle). I have simply taken it for granted that we have an intuitive understanding of what it means to be self-determining, and of why that is thought to be an important value. But we need to examine this issue more closely, for communitarians challenge many of the standard assumptions about the nature and value of self-determination. In particular, communitarians argue that liberals both misconstrue our capacity for self-determination, and neglect the social preconditions under which that capacity can be meaningfully exercised. I will look at these two objections in turn, after first laying out the liberal account of the value of self-determination.

Many liberals think that the value of self-determination is so obvious that it does not require any defence. Allowing people to be self-determining is, they say, the only way to respect them as fully moral beings. To deny self-determination is to treat someone like a child or an animal, rather than a full member of the community.

But this is too quick. We know that some people are not well equipped to deal with the difficult decisions life requires. They make mistakes about their lives, choosing to do trivial, degrading, even harmful, things. If we are supposed to show concern for people, why should we not stop people from making such mistakes? When people are unable to deal effectively with life, respecting their self-determination may amount in practice to abandoning them to an unhappy fate. Saying that we ought to respect people's self-determination under these circumstances becomes an expression of indifference rather than concern. Dworkin says that it is 'the final evil of a genuinely unequal distribution of resources' that some people 'have been cheated of the chance others have had to make something valuable of their lives' (Dworkin 1981: 219). But what about those who are unable to make something valuable of their lives even when they have the chance? Do we not have obligations here too?

Liberals do, of course, leave room in their theory for acts of paternalism—for example, in our relations with children, the demented, and the otherwise temporarily incapacitated.[1] But liberals insist that every competent adult be provided with a sphere of self-determination which must be respected by others. As Mill put it, it is the right and prerogative of each person, once they have reached the maturity of their years, to interpret for themselves the meaning and value of their experiences. For those who pass the threshold of age and mental competence, the right to be self-determining in the major decisions in life is inviolate.

But why should we view self-determination in terms of such a threshold? Some of the people who are beyond the 'age of reason', and who have a mental competence above the agreed-upon minimum, still make poor choices about how to lead their lives. Being a 'competent adult', in the sense that one is not mentally handicapped, is no guarantee that one is good at making something valuable of one's life. So why should government not decide what sort of lives are best for its citizens?

Marxist perfectionism is one example of such a policy, for it prohibits people from making what it views as a bad choice—i.e. choosing to engage in alienated labour. I argued that this policy is unattractive, for it relies on too narrow an account of the good. It identifies our good with a single activity—productive labour—on the grounds that it alone makes us distinctively human. But not all paternalist or perfectionist policies are based on such an implausible account of the good life. Consider a policy which subsidizes theatre while taxing professional wrestling. Defenders of such a policy need not say that theatre is the only, or even the most important, good in life. They simply claim that of these two options, as they currently exist, theatre is the more valuable. They may have a number of different arguments for this claim. Studies may reveal that theatre is stimulating, whereas wrestling produces frustration and docility; or that wrestling-fans often come to regret their past activities, whereas theatre-goers rarely regret theirs; or that the majority of people who have tried both forms of entertainment prefer theatre. Under these circumstances, the claim that theatre is better entertainment than wrestling has some plausibility. Why then should the government not encourage people to attend the theatre, and save them from wasting their lives on wrestling?

Liberals view such policies, no matter how plausible the underlying theory of the good, as an illegitimate restriction on self-determination. If there are willing participants and spectators for wrestling, then the anti-wrestling policy is an unjustified restriction on people's freely chosen leisure. How can liberals defend the importance of people being free to choose their own leisure? Since the argument for perfectionism depends on the assumption that people can make mistakes about the value of their activities, one possible line of defence is to deny that people can be mistaken in their judgements of what is valuable in life. Defenders of self-determination might argue that judgements of value, unlike judgements of fact, are simply the expressions of our subjective likes and dislikes. These choices are ultimately arbitrary, incapable of rational justification or criticism. All such choices are equally rational, and so the state has no reason to interfere in them. Many perfectionists have assumed that this sort of scepticism about value judgements has to be the liberal position, for if we concede the possibility of people making mistakes, then surely the government must encourage the right ways of life, and discourage or prohibit mistaken ones (Unger 1984: 52, 66–7; Jaggar 1983: 194, 174; Sullivan 1982: 38–40).

But liberals do not endorse scepticism. One reason is that

scepticism does not in fact support self-determination. If people cannot make mistakes in their choices, then neither can governments. If all ways of life are equally valuable, then no one can complain when the government chooses a particular way of life for the community. Hence scepticism leaves the issue unresolved.

How do liberals defend the importance of self-determination? We need to look more closely at this idea. Self-determination involves deciding what to do with our lives. How do we make such decisions? At the most general level, our aim is to lead a good life, to have those things that a good life contains. Put at such a general level, that claim may seem quite uninformative. But it has important consequences. For, as we saw in chapter 2, leading a good life is different from leading the life we currently believe to be good (ch. 2, s. 2c above). We recognize that we may be mistaken about the value of our current activities. We may come to see that we have been wasting our lives, pursuing trivial goals that we had mistakenly considered of great importance. This is the stuff of great novels—the crisis in faith. But the assumption that this could happen to all of us, and not just to the tragic heroine, is needed to make sense of the way we deliberate about important decisions in our life. We deliberate carefully because we know we could make wrong decisions. And not just in the sense of predicting wrongly, or of calculating uncertainties. For we deliberate even when we know what will happen, and we may regret our decisions even when things went as planned. I may succeed in becoming the best pushpin player in the world, but then come to realize that pushpin is not as valuable as poetry, and regret that I had ever embarked on that project.

Deliberation, then, does not only take the form of asking which course of action maximizes a particular value that is held unquestioned. We also question, and worry about, whether that value is really worth pursuing. As Rawls puts it,

As free persons, citizens recognize one another as having the moral power to have a conception of the good. This means that they do not view themselves as inevitably tied to the pursuit of the particular conception of the good and its final ends which they espouse at any given time. Instead, as citizens, they are regarded as, in general, capable of revising and changing this conception on reasonable and rational grounds. Thus it is held to be permissible for citizens to stand apart from conceptions of the good and to survey and assess their various final ends. (Rawls 1980: 544)

We can 'stand apart' from our current ends, and question their value to us. The concern with which we make these judgements, at certain points in our lives, only makes sense on the assumption that our essential interest is in living a good life, not the life we currently believe to be good. We do not just make such judgements, we worry, sometimes agonize, over them—it is important to us that we do not lead our lives on the basis of false beliefs about the value of our activities (Raz 1986: 300–2).

The idea that some things really are worth doing, and others are not, goes very deep in our self-understanding. We take seriously the distinction between worthwhile and trivial activities, even if we are not always sure which things are which. Self-determination is, to a large extent, the task of making these difficult, and potentially fallible, judgements, and our political theory should take this difficulty and fallibility into account.

Should we therefore be perfectionists, supporting state policies which discourage trivial activities to which people are mistakenly attached? Not necessarily. For one thing, no one may be in a better position than I am to know my own good. Even if I am not always right, I may be more likely to be right than anyone else. Mill defended a version of this argument by claiming that each person contained a unique personality whose good was different from that of anyone else. The experience of others, therefore, provides no grounds for overriding my judgement. This is the opposite of Marxist perfectionism—where Marxists say that each person's good lies in a capacity she shares with all other humans, Mill says it lies in something she shares with no one else. But surely both extremes are wrong. Our good is neither universal nor unique, but tied in important ways to the cultural practices we share with others in our community. We share enough with others around us that a well-intentioned perfectionist government could, by drawing on the wisdom and experience of others, arrive at a reasonable set of beliefs about its citizens' good. Of course, we might doubt that governments have either the right intentions or abilities to execute such a programme. But nothing in principle excludes the possibility that governments can identify mistakes in people's conceptions of the good.

Why then do liberals oppose state paternalism? Because, they argue, no life goes better by being led from the outside according to values the person does not endorse. My life only goes better if I am leading it from the inside, according to my beliefs about value. Praying to God may be a valuable activity, but I have to believe that it is a worthwhile thing to do—that it has some worthwhile point. We can coerce someone into going to church and making the right physical movements, but we will not make her life better that way. It will not work, even if the coerced person is mistaken in her belief that praying to God is a waste of time, because a valuable life has to be led from the inside.[2] A perfectionist policy that violates this 'endorsement constraint', by trying to bypass or override people's beliefs about values, is self-defeating (Dworkin 1989: 486–7). It may succeed in getting people to pursue valuable activities, but it does so under conditions in which the activities cease to have value for the individuals involved. If I do not see the point of an activity, then I will gain nothing from it. Hence paternalism creates the very sort of pointless activity that it was designed to prevent.

So we have two preconditions for the fulfilment of our essential interest in leading a life that is good. One is that we lead our life from the inside, in accordance with our beliefs about what gives value to life; the other is that we be free to question those beliefs, to examine them in the light of whatever information, examples, and arguments our culture can provide. People must therefore have the resources and liberties needed to lead their lives in accordance with their beliefs about value, without being penalized for unorthodox religious or sexual practices, etc. Hence the traditional liberal concern for civil and personal liberties. And individuals must have the cultural conditions necessary to acquire an awareness of different views about the good life, and to acquire an ability to examine these views intelligently. Hence the traditional liberal concern for education, freedom of expression, freedom of the press, artistic freedom, etc. These liberties enable us to judge what is valuable in life in the only way we can judge such things—i.e. by exploring different aspects of our shared cultural heritage.

This account of the value of self-determination forms the basis of Rawls's liberty principle. According to Rawls, freedom of choice is needed precisely to find out what is valuable in life—to form, examine, and revise our beliefs about value.[3] Liberty helps us come to know our good, to 'track bestness', in Nozick's phrase (Nozick 1981: 314, 410–11, 436–40, 498–504; cf. Dworkin 1983: 24–30). Since we have an essential interest in getting these beliefs right and acting on them, government treats people with equal concern and respect by providing each person with the liberties and resources needed to examine and act on these beliefs.

Rawls argues that this account of self-determination should lead us to endorse a 'neutral state'—i.e. a state which does not justify its actions on the basis of the intrinsic superiority or inferiority of conceptions of the good life, and which does not deliberately attempt to influence people's judgements of the value of these different conceptions.[4] He contrasts this with perfectionist theories which include a particular view, or range of views, as to what attributes are most worth developing. Perfectionists demand that resources should be distributed so as to encourage such development. What one gets depends on how much one needs to pursue, or how much one contributes to, this preferred view of the good life. People are not, therefore, free to choose their own conception of the good life, at least not without being penalized by society. People make mistakes about the good life, and the state has the responsibility to teach its citizens about a virtuous life. It abandons that responsibility to its citizens if it funds, or perhaps even tolerates, life-plans that have misconceived views about human excellence.

For Rawls, on the other hand, our essential interests are harmed by attempts to enforce a particular view of the good life on people. He favours the distribution of primary goods, based on a 'thin theory of the good', which can be used to advance many different ways of life (cf. ch. 3, s. 3 above). If we only have access to resources that are useful for one plan of life, then we will be unable to act on our beliefs about value, should we come to believe that that one preferred conception of the good life is misguided. (Or, at any rate, we will be unable to do so without suffering some penalty in social benefits.) Since lives have to be led from the inside, someone's essential interest in leading a life that is good is not advanced when society penalizes, or discriminates against, the projects that she, on reflection, feels are most valuable for her. Distributing resources according to a 'thin theory of the good', or what Dworkin calls 'resources in the widest sense', best enables people to act on and examine their beliefs about value, and that is the most appropriate way to promote people's essential interest in leading a good life.

2. COMMUNITARIANISM AND THE COMMON GOOD

Communitarians object to the neutral state. They believe it should be abandoned for a 'politics of the common good' (Sandel 1984*b*: 16–17; Taylor 1986). This contrast between the 'politics of neutrality' and communitarianism's 'politics of the common good' can be misleading. There is a 'common good' present in liberal politics as well, since the policies of a liberal state aim at promoting the interests of the members of the community. The political and economic processes by which individual preferences are combined into a social choice function are liberal modes of determining the common good. To affirm state neutrality, therefore, is not to reject the idea of a common good, but rather to provide an interpretation of it (Holmes 1989: 239–40). In a liberal society, the common good is the result of a process of combining preferences, all of which are counted equally (if consistent with the principles of justice). All preferences have equal weight 'not in the sense that there is an agreed public measure of intrinsic value or satisfaction with respect to which all these conceptions come out equal, but in the sense that they are not evaluated at all from a [public] standpoint' (Rawls 1982*b*: 172). As we have seen, this anti-perfectionist insistence on state neutrality reflects the belief that people's interest in leading a good life is not advanced when society discriminates against the projects that they believe are most valuable for them. Hence the common good in a liberal society is adjusted to fit the pattern of preferences and conceptions of the good held by individuals.

In a communitarian society, however, the common good is conceived of as a substantive conception of the good life which defines the community's 'way of life'. This common good, rather than adjusting itself to the pattern of people's preferences, provides a standard by which those preferences are evaluated. The community's way of life forms the basis for a public ranking of conceptions of the good, and the weight given to an individual's preferences depends on how much she conforms or contributes to this common good. The public pursuit of the shared ends which define the community's way of life is not, therefore, constrained by the requirement of neutrality. It takes precedence over the claim of individuals to the resources and liberties needed to pursue their own conceptions of the good. A communitarian state can and should encourage people to adopt conceptions of the good that conform to the community's way of life, while discouraging conceptions of the good that conflict with it. A communitarian state is, therefore, a perfectionist state, since it involves a public ranking of the value of different ways of life. But whereas Marxist perfectionism ranks ways of life according to a trans-historical account of the human good, communitarianism ranks them according to their conformity to existing practices.

Why should we prefer this 'politics of the common good' over liberal neutrality? Liberals say that state neutrality is required to respect people's self-determination. Communitarians, however, object both to the liberal idea of self-determination, and to the supposed connection between self-determination and neutrality. I will consider these two objections in turn.

3. THE UNENCUMBERED SELF

On the liberal view of the self, individuals are considered free to question their participation in existing social practices, and opt out of them, should those practices seem no longer worth pursuing. As a result, individuals are not defined by their membership in any particular economic, religious, sexual, or recreational relationship, since they are free to question and reject any particular relationship. Rawls summarizes this liberal view by saying that 'the self is prior to the ends which are affirmed by it' (Rawls 1971: 560), by which he means that we can always step back from any particular project and question whether we want to continue pursuing it. No end is exempt from possible revision by the self. This is often called the 'Kantian' view of the self, for Kant was one of the strongest defenders of the view that the self is prior to its socially given roles and relationships, and is free only if it is capable of holding these features of its social situation at a distance and judging them according to the dictates of reason (Taylor 1979: 75–8, 132–3).

Communitarians believe that this is a false view of the self. It ignores the fact that the self is 'embedded' or 'situated' in existing social practices, that we cannot always stand back and opt out of them. Our social roles and relationships, or at least some of them, must be taken as givens for the purposes of personal deliberation. As MacIntyre puts it, in deciding how to lead our lives, we 'all approach our own circumstances as bearers of a particular social identity. . . . Hence what is good for me has to be the good for one who inhabits these roles' (MacIntyre 1981: 204–5). Self-determination, therefore, is exercised within these social roles, rather than by standing outside of them. And so the state respects our self-determination not by enabling us to stand back from our social roles, but by encouraging a deeper immersion in and understanding of them, as the politics of the common good seeks to accomplish.

Communitarians have a number of different arguments against the liberal account of the self and its ends. I will consider three, which can be summarized this way: the liberal view of the self (1) is empty; (2) violates our self-perceptions; and (3) ignores our embeddedness in communal practices.[5]

Firstly, the emptiness argument. Being free to question all our social roles is self-defeating, Charles Taylor says, because 'complete freedom would be a void in which nothing would be worth doing, nothing would deserve to count for anything. The self which has arrived at freedom by setting aside all external obstacles and impingements is characterless, and hence without defined purpose' (Taylor 1979: 157). True freedom must be 'situated', Taylor argues. The desire to subject all aspects of our social situation to our rational self-determination is empty, because the demand to be self-determining is indeterminate. It 'cannot specify any content to our action outside of a situation which sets goals for us, which thus imparts a shape to rationality and provides an inspiration for creativity' (Taylor 1979: 157). We must accept the goal that our situation 'sets for us'—if we do not, then the quest for self-determination leads to Nietzschean nihilism, the rejection of all communal values as ultimately arbitrary: 'One after the other, the authoritative horizons of life, Christian and humanist, are cast off as shackles on the will. Only the will to power remains' (Taylor 1979: 159). If we deny that communal values are 'authoritative horizons', then they will appear as arbitrary limits on our will, and hence our freedom will require rejecting them all (MacIntyre 1981: ch. 9).

But this misconstrues the role that freedom plays in liberal theories. According to Taylor, liberals claim that the freedom to choose our projects is inherently valuable, something to be pursued for its own sake, a claim that Taylor rejects as empty. Instead, he says, there has to be some project that is worth pursuing, some task that is worth fulfilling. But the concern for freedom within liberalism does not take the place of these tasks and projects. On the contrary, the liberal defence of freedom rests precisely on the importance of those projects. Liberals do not say that we should have the freedom to select our projects for its own sake, because freedom is the most valuable thing in the world. Rather, our projects and tasks are the most important things in our lives, and it is because they are so important that we should be free to revise them, should we come to believe that they are not worthwhile. Our projects are the most important things in our lives, but since our lives have to be led from the inside, in accordance with our beliefs about value, we should be free to form, revise, and act on our plans of life. Freedom of choice is not pursued for its own sake, but as a precondition for pursuing those projects that are valued for their own sake.

Some liberals have endorsed the position Taylor rightly criticizes as empty. Isaiah Berlin attributes it to Mill, for example (Berlin 1969: 192; but cf. Ladenson 1983: 149–53). Claiming that freedom of choice is intrinsically valuable may seem like an effective way of defending a broad range of liberal freedoms. But the implications of that claim conflict with the way we understand the value in our lives in at least two important ways: (1) Saying that freedom of choice is intrinsically valuable suggests that the more we exercise our capacity for choice, the more free we are, and hence the more valuable our lives are. But that is false, and indeed perverse. It quickly leads to the existentialist view that we should wake up each morning and decide anew what sort of person we should be. This is perverse because a valuable life is a life filled with commitments and relationships. These give our lives depth and character. And what makes them commitments is precisely that they are not the sort of thing that we question every day. We do not suppose that someone who makes twenty marriage choices is in any way leading a more valuable life than someone who has no reason to question her original choice. A life with more marital choices is not even *ceteris paribus* better than a life with fewer such choices. (2) Saying that freedom of choice is intrinsically valuable suggests that the value we seek in our actions is freedom, not the value internal to the activity itself. This suggestion is endorsed by Carol Gould. She says that while we seem to act for the sake of the purposes internal to a given project, truly free activity has freedom itself as the ultimate end: 'Thus freedom is not only the activity that creates value but is that for the sake of which all these other values are pursued and therefore that with respect to which they become valuable' (Gould 1978: 118).

But this is false. Firstly, as Taylor notes, telling people to act freely does not tell them what particular actions are worth doing. But even if it provided determinate guidance, it still presents a false view of our motivations. If I am writing a book, for example, my motivation is not to be free, but to say something that is worth saying. Indeed, if I did not really want to say anything, except in so far as it is a way of being free, then my writing would not be fulfilling. What and how I write would become the result of arbitrary and ultimately unsatisfying choices. If writing is to be intrinsically valuable, I have to care about what I am saying, I have to believe that writing is worth doing for its own sake. If we are to understand the value people see in their projects, we have to look to the ends which are internal to them. I do not pursue my writing for the sake of my freedom. On the contrary, I pursue my writing for its own sake, because there are things which are worth saying. Freedom is valuable because it allows me to say them.

The best defence of individual freedoms is not necessarily the most direct one, but the one which best accords with the way that people on reflection understand the value of their lives. And if we look at the value of freedom in this way, then it seems that freedom of choice, while central to a valuable life, is not the value which is centrally pursued in such a life.

No one disagrees that projects have to be our primary concern— that does not distinguish the liberal and the communitarian. The real debate is not over whether we need such tasks, but over how we acquire them and judge their worth. Taylor seems to believe that we can acquire these tasks only by treating communal values as 'authoritative horizons' which 'set goals for us' (Taylor 1979: 157–9). Liberals, on the other hand, insist that we have an ability to detach ourselves from any particular social practice. No particular task is set for us by society, no particular practice has authority that is beyond individual judgement and possible rejection. We can and should acquire our tasks through freely made personal judgements about the cultural structure, the matrix of understandings and alternatives passed down to us by previous generations, which offers us possibilities we can either affirm or reject. Nothing is 'set for us'; nothing is authoritative before our judgement of its value.

Of course, in making that judgement, we must take something as a 'given'—we ask what is good for us now, given our place in school, work, or family. Someone who is nothing but a free rational being would have no reason to choose one way of life over another (Sandel 1982: 161–5; Taylor 1979: 157; Crowley 1987: 204–5). But liberals believe that what we put in 'the given' in order to make meaningful judgements can not only be different between individuals but also can change within one individual's life. If at one time we make choices about what is valuable given our commitment to a certain religious life, we could later come to question that commitment, and ask what is valuable given our commitment to our family. The question then is not whether we must take something as given in making judgements about the value of our activity. Rather, the question is whether an individual can question and possibly replace what is in 'the given', or whether the given has to be set for us by the community's values. Taylor fails to show that we must take communal values as given, that it is empty to say that such communal values should be subject to individual evaluation and possible rejection.

One can weaken the communitarian objection by arguing that even if we can get our purposes this way, unset by the community, we none the less should treat communal ends as authoritative. We should do this because the liberal view relies on a false account of the self. The liberal view, we have seen, is that 'the self is prior to its ends', in the sense that we reserve the right to question even our most deeply held convictions about the nature of the good life. Michael Sandel, however, argues that the self is not prior to, but rather constituted by, its ends—we cannot distinguish 'me' from 'my ends'. Our selves are at least partly constituted by ends that we do not choose, but rather discover by virtue of our being embedded in some shared social context (Sandel 1982: 55–9, 152–4). Since we have these constitutive ends, our lives go better not by having the conditions needed to select and revise our projects, but by having the conditions needed to come to an awareness of these shared constitutive ends. A politics of the common good, by expressing these shared constitutive ends, enables us to 'know a good in common that we cannot know alone' (Sandel 1982: 183).

Sandel has two arguments for this claim, which I will call the 'self-perception' and 'embedded-self' arguments. The first argument goes like this: Rawls's view of the 'unencumbered self' does not correspond with our 'deepest self-understanding' in the sense of our deepest self-perception. According to Sandel, if the self is prior to its ends, then we should, when introspecting, be able to see through our particular ends to an unencumbered self. But, Sandel notes, we do not perceive our selves as being unencumbered: Rawls's view of the self as 'given prior to its ends, a pure subject of agency and possession, ultimately thin', is 'radically at odds with our more familiar notion of ourselves as beings "thick with particular traits"' (Sandel 1982: 94, 100). On Rawls's view, 'to identify any characteristics as *my* aims, ambitions, desires, and so on, is always to imply some subject "me" standing behind them, at a certain distance' (Sandel 1984*a*: 86). There would have to be this thing, a self, which has some shape, albeit an ultimately thin shape, standing at some distance behind our ends. To accept Rawls, I would have to see myself as this propertyless thing, a disembodied rather ghostly object in space, or as Rorty puts it, as a kind of 'substrate' lying 'behind' my ends (Rorty 1985: 217). In contrast, Sandel says that our deepest self-perceptions always include some motivations, which shows that some ends are constitutive of the self.

But the question of perception here is misleading. What is central to the liberal view is not that we can perceive a self prior to its ends, but that we understand ourselves to be prior to our ends in the sense that no end or goal is exempt from possible re-examination. For re-examination to be meaningfully conducted, I must be able to see my self encumbered with different motivations from those I now have, in order that I have some reason to choose one over another as more valuable for me. My self is, in this sense, perceived prior to its ends, i.e. I can always envisage my self without its *present* ends. But this does not require that I can ever perceive a self unencumbered by any ends—the process of practical reasoning is always one of comparing one 'encumbered' potential self with another 'encumbered' potential self. There must always be some ends given with the self when we engage in such reasoning, but it does not follow that any particular ends must always be taken as given with the self. As I said before, it seems that what is given with the self can change over the course of a lifetime. Thus there is a further claim that Sandel must establish: he must show not only that we cannot perceive a totally unencumbered self, but that we cannot perceive our self encumbered by a different set of ends. This requires a different argument, which I call the embedded-self argument.

This third argument contrasts the communitarian view of practical reasoning as self-discovery with the liberal view of practical reasoning as judgement. For liberals, the question about the good life requires us to make a judgement about what sort of a person we wish to be or become. For communitarians, however, the question requires us to discover who we already are. For communitarians, the relevant question is not 'What should I be, what sort of life should I lead?' but 'Who am I?'. The self 'comes by' its ends not 'by choice' but 'by discovery', not 'by choosing that which is already given (this would be unintelligible) but by reflecting on itself and inquiring into its constituent nature, discerning its laws and imperatives, and acknowledging its purposes as its own' (Sandel 1982: 58). For example, Sandel criticizes Rawls's account of community, because 'while Rawls allows that the good of community can be internal to the extent of engaging the aims and values of the self, it cannot be so thoroughgoing as to reach beyond the motivations to the subject of motivations' (Sandel 1982: 149). On a more adequate account, Sandel claims, communal values are not just affirmed by the members of the community, but define their identity. The shared pursuit of a communal goal is 'not a relationship they choose (as in a voluntary association) but an attachment they discover, not merely an attribute but a constituent of their identity' (Sandel 1982: 150). The good for such members is found by a process of self-discovery—by achieving awareness of, and acknowledging the claims of, the various attachments they 'find'.

But surely it is Sandel here who is violating our deepest self-understandings. For we do not think that this self-discovery replaces or forecloses judgements about how to lead our life. We do not consider ourselves trapped by our present attachments, incapable of judging the worth of the goals we inherited or ourselves chose earlier. We do indeed find ourselves in various relationships, but we do not always like what we find. No matter how deeply implicated we find ourselves in a social practice, we feel capable of questioning whether the practice is a valuable one—a questioning which is not meaningful on Sandel's account. (How can it not be valuable since the good for me just is coming to a greater self-awareness of the attachments I find myself in?) The idea that deliberation is completed by this process of self-discovery (rather than by judgements of the value of the attachments we discover) seems pretty facile.

In places, Sandel admits that practical reasoning is not just a question of self-discovery. He says that the boundaries of the self, although constituted by its ends, are none the less flexible and can be redrawn, incorporating new ends and excluding others. In his words, 'the subject is empowered to participate in the constitution of its identity'; on his account 'the bounds of the self [are] open and the identity of the subject [is] the product rather than the premise of its agency' (Sandel 1982: 152). The subject can, after all, make choices about which of the 'possible purposes and ends, all impinging indiscriminately on its identity' it will pursue, and which it will not (Sandel 1982: 152). The self, constituted by its ends, can be 'reconstituted' as it were, so self-discovery is not enough. But at this point it is not clear whether the distinction between the two views does not collapse.

There are apparent differences here. Sandel claims that the self is constituted by its ends, and that the boundaries of the self are fluid, whereas Rawls says that the self is prior to its ends, and its boundaries are fixed antecedently. But these two differences hide a more fundamental identity; both accept that the *person* is prior to her ends. They disagree over where, within the person, to draw the boundaries of the 'self'; but this question, if it is indeed a meaningful question, is one for the philosophy of mind, not political philosophy. For so long as Sandel admits that the person can re-examine her ends—even the ends constitutive of her 'self'— then he has failed to justify communitarian politics. He has failed to show why individuals should not be given the conditions appropriate to that re-examining, as an indispensable part of leading the best possible life. And amongst these conditions should be the liberal guarantees of personal independence necessary to make the judgement freely. Sandel trades on an ambiguity in the view of the person that he uses in defending communitarian politics. The strong claim (that self-discovery replaces judgement) is implausible, and the weak claim (which allows that a self constituted by its ends can none the less be reconstituted), while attractive, fails to distinguish him from the liberal view.[6]

Sandel says that liberalism ignores the way we are embedded in our social roles. He emphasizes that as 'self-interpreting beings', we can interpret the meaning of these constitutive attachments (Sandel 1984a: 91). But the question is whether we can reject them entirely should we come to view them as trivial or degrading. On one interpretation of communitarianism, we cannot, or at any rate, we should not. On this view, we neither choose nor reject these attachments, rather we find ourselves in them. Our goals come not by choice, but by self-discovery. A Christian housewife in a monogamous heterosexual marriage can interpret what it means to be a Christian or a housewife—she can interpret the meaning of these shared religious, economic, and sexual practices. But she cannot stand back and decide that she does not want to be a Christian at all, or a housewife. I can interpret the meaning of the roles I find myself in, but I cannot reject the roles themselves, or the goals internal to them, as worthless. Since these goals are constitutive of me as a person, they have to be taken as given in deciding what to do with my life; the question of the good in my life can only be a question of how best to interpret their meaning. It makes no sense to say that they have no value for me, since there is no 'me' standing behind them, no self prior to these constitutive attachments.

It is unclear which if any communitarians hold this view consistently. It is not a plausible position, since we can and do make sense of questions not just about the meaning of the roles we find ourselves in, but also about their value. Perhaps communitarians do not mean to deny that; perhaps their idea of our embeddedness is not incompatible with our rejecting the attachments we find ourselves in. But then the advertised contrast with the liberal view is a deception, for the sense in which communitarians view us as embedded in communal roles incorporates the sense in which liberals view us as independent of them, and the sense in which communitarians view practical reasoning as a process of 'self-discovery' incorporates the sense in which liberals view it as a process of judgement and choice. The differences would be merely semantic. And once we agree that individuals are capable of questioning and rejecting the value of the community's way of life, then the attempt to discourage such questioning through a 'politics of the common good' seems an unjustified restriction on people's self-determination.

4. THE SOCIAL THESIS

Many communitarians criticize liberalism, not for its account of the self and its interests, but for neglecting the social conditions required for the effective fulfilment of those interests. For example, Taylor claims that many liberal theories are based on 'atomism', on an 'utterly facile moral pyschology' according to which individuals are self-sufficient outside of society. Individuals, according to atomistic theories, are not in need of any communal context in order to develop and exercise their capacity for self-determination. Taylor argues instead for the 'social thesis', which says that this capacity can only be exercised in a certain kind of society, with a certain kind of social environment (Taylor 1985: 190–1; cf. Jaggar 1983: 42–3; Wolgast 1987: ch. 1).

If this really were the debate, then we would have to agree with the communitarians, for the 'social thesis' is clearly true. The view that we might exercise the capacity for self-determination outside of society is absurd. But liberals like Rawls and Dworkin do not deny the social thesis. They recognize that individual autonomy cannot exist outside a social environment that provides meaningful choices and that supports the development of the capacity to choose amongst them (e.g. Rawls 1971: 563–4; Dworkin 1985: 230–3).

Taylor believes, however, that the social thesis requires us to abandon liberal neutrality, for a neutral state cannot adequately protect the social environment necessary for self-determination.[7] The social thesis tells us that the capacity to choose a conception of the good can only be exercised in a particular sort of community, and, Taylor argues, this sort of community can only be sustained by a politics of the common good. In other words, some limits on self-determination are required to preserve the social conditions which enable self-determination. I will consider three versions of this claim: one about the need to sustain a cultural structure that provides people with meaningful options; a second about the need for shared forums in which to evaluate these options; and a third about the preconditions for political legitimacy. In each case, communitarians invoke the social thesis to show how a concern for self-determination supports, rather than precludes, communitarian politics.

(a) Duties to protect the cultural structure

Meaningful choices concerning our projects require meaningful options, and (the social thesis tells us) these options come from our culture. Liberal neutrality, however, is incapable of ensuring the existence of a rich and diverse culture which provides such options. Self-determination requires pluralism, in the sense of a diversity of possible ways of life, but

Any collective attempt by a liberal state to protect pluralism would itself be in breach of liberal principles of justice. The state is not entitled to interfere in the movement of the cultural market place except, of course, to ensure that each individual has a just share of available necessary means to exercise his or her moral powers. The welfare or demise of particular conceptions of the good and, therefore, the welfare or demise of social unions of a particular character is not the business of the state. (Cragg 1986: 47)

Liberals believe that a state which intervenes in the cultural marketplace to encourage or discourage any particular way of life restricts people's self-determination. However, if the cultural marketplace proceeds on its own it will eventually undermine the cultural structure which supports pluralism. As Joseph Raz puts it, 'Supporting valuable forms of life is a social rather than an individual matter . . . perfectionist ideals require public action for their viability. Anti-perfectionism in practice would lead not merely to a political stand-off from support for valuable conceptions of the good. It would undermine the chances of survival of many cherished aspects of our culture' (Raz 1986: 162). Liberal neutrality is therefore self-defeating.

This is an important objection. Many liberals are surprisingly silent about the possibility that 'the essential cultural activities which make a great diversity conceivable to people [could] begin to falter'. As Taylor says, 'it is as though the conditions of a creative, diversifying freedom were given by nature' (Taylor 1985: 206 n. 7). Rawls attempts to answer this worry by claiming that good ways of life will in fact sustain themselves in the cultural marketplace without state assistance, because in conditions of freedom, people are able to recognize the worth of good ways of life, and will support them (Rawls 1971: 331–2; cf. Waldron 1989: 1138). But this is inadequate. The interests people have in a good way of life, and the forms of support they will voluntarily provide, do not necessarily involve sustaining its existence for future generations. My interest in a valuable social practice may be best promoted by depleting the resources which the practice requires to survive beyond my lifetime. Consider the preservation of historical artefacts and sites, or of natural wilderness areas. The wear and tear caused by the everyday use of these things would prevent future generations from experiencing them, were it not for state protection. So even if the cultural marketplace can be relied on to ensure that existing people can identify valuable ways of life, it cannot be relied on to ensure that future people have a valuable range of options.

So let us grant Raz's argument that state support may be needed to ensure the survival of an adequate range of options for those who have not yet formed their aims in life. Why does that require rejecting neutrality? Consider two possible cultural policies: In the first case, the government ensures an adequate range of options by providing tax credits to individuals who make culture-supporting contributions in accordance with their personal perfectionist ideals. The state acts to ensure that there is an adequate range of options, but the evaluation of these options occurs in civil society, outside the apparatus of the state (cf. Dworkin 1985: ch. 11). In the second case, the evaluation of different conceptions of the good becomes a political question, and the government intervenes, not simply to ensure an adequate range of options, but to promote particular options. Raz's argument shows that one or other of these policies must be implemented, but he has not given a decisive reason, or any reason at all, to prefer one policy over the other.

Hence the existence of duties concerning the protection of the cultural structure is not incompatible with neutrality. In fact, Dworkin emphasizes our duty to protect the cultural structure from 'debasement or decay' (Dworkin 1985: 230). Like Taylor, he talks about how the capacity imaginatively to conceive conceptions of the good life requires specialized debate among intellectuals who attempt to define and clarify the alternatives facing us, or the presence of people who attempt to bring the culture of the past to life again in the art of the present, or who sustain the drive to cultural innovation, and about how the state can and should protect these essential cultural activities (Taylor 1985: 204–6; Dworkin 1985: 229–32). And while Rawls does not include state support for culture in his theory of justice, since he thinks that the operation of the two principles would in fact protect these preconditions of a diverse culture, there is no reason why he should reject such support where this is not the case (Rawls 1971: 331, 441–2, 522–9). Like Dworkin, he would simply insist that it is not the job of the state to rank the relative value of the various options within the culture.

A communitarian state might hope to improve the quality of people's options, by encouraging the replacement of less valuable aspects of the community's ways of life by more valuable ones. But liberal neutrality also hopes to improve the range of people's options. Freedom of speech and association allows each group to pursue and advertise its way of life, and those ways of life that are unworthy will have difficulty attracting adherents. Since individuals are free to choose between competing visions of the good life, liberal neutrality creates a marketplace of ideas, as it were, and how well a way of life does in this market depends on the goods it can offer to prospective adherents. Hence, under conditions of freedom, satisfying and valuable ways of life will tend to drive out those which are unsatisfying. Liberals endorse civil liberties in part precisely because they make it possible 'that the worth of different modes of life should be proved practically' (Mill 1974: 54).

Liberals and communitarians both aim to secure the range of options from which individuals make their autonomous choices. What they disagree on is where perfectionist ideals should be invoked. Are good ways of life more likely to establish their greater worth when they are evaluated in the cultural marketplace of civil society, or when the preferability of different ways of life is made a matter of political advocacy and state action? Hence the dispute should perhaps be seen as a choice, not between perfectionism and neutrality, but between social perfectionism and state perfectionism—for the flip side of state neutrality is support for the role of perfectionist ideals in civil society.

(b) Neutrality and collective deliberations

Some communitarians argue that the liberal preference for the cultural marketplace over the state as the appropriate arena for evaluating different ways of life stems from an atomistic belief that judgements about the good are only autonomous when they are made by isolated individuals who are protected from social pressure. Liberals think that autonomy is promoted when judgements about the good are taken out of the political realm. But in reality individual judgements require the sharing of experiences and the give and take of collective deliberation. Individual judgements about the good depend on the collective evaluation of shared practices. They become a matter of subjective and arbitrary whim if they are cut off from collective deliberations:

> [S]elf-fulfillment and even the working out of personal identity and a sense of orientation in the world depend upon a communal enterprise. This shared process is the civic life, and its root is involvement with others: other generations, other sorts of persons whose differences are significant because they contribute to the whole upon which our particular sense of self depends. Thus mutual interdependency is the foundational notion of citizenship . . . Outside a linguistic community of shared practices, there would be biological *homo sapiens* as logical abstraction, but there could not be human beings. This is the meaning of the Greek and medieval dictum that the political community is ontologically *prior* to the individual. The *polis* is, literally, that which makes man, as human being, possible. (Sullivan 1982: 158, 173)

Or, as Crowley puts it, state perfectionism is

> an affirmation of the notion that men living in a community of shared experiences and language is the only context in which the individual and society can discover and test their values through the essentially political activities of discussion, criticism, example, and emulation. It is through the existence of organised public spaces, in which men offer and test ideas against one another . . . that men come to understand a part of who they are. (Crowley 1987: 282; cf. Beiner 1983: 152)

The state is the proper arena in which to formulate our visions of the good, because these visions require shared enquiry. They cannot be pursued, or even known, by solitary individuals.

But this misconstrues the sense in which Rawls claims that the evaluation of ways of life should not be a public concern. Liberal neutrality does not restrict the scope of perfectionist ideals in the collective activities of individuals and groups. Collective activity and shared experiences concerning the good are at the heart of the 'free internal life of the various communities of interests in which persons and groups seek to achieve, in modes of social union consistent with equal liberty, the ends and excellences to which they are drawn'. Rawls's argument for the priority of liberty is grounded in the importance of this 'free social union with others' (Rawls 1971: 543). He simply denies that 'the coercive apparatus of the state' is an appropriate forum for those deliberations and experiences:

While justice as fairness allows that in a well-ordered society the values of excellence are recognized, the human perfections are to be pursued within the limits of the principle of free association. . . . [Persons] do not use the coercive apparatus of the state to win for themselves a greater liberty or larger distributive shares on the grounds that their activities are of more intrinsic value. (Rawls 1971: 328–9)

Unfortunately, communitarians rarely distinguish between collective activities and political activities. It is of course true that participation in shared linguistic and cultural practices is what enables individuals to make intelligent decisions about the good life. But why should such participation be organized through the state, rather than through the free association of individuals? It is true that we should 'create opportunities for men to give voice to what they have discovered about themselves and the world and to persuade others of its worth' (Crowley 1987: 295). But a liberal society does create opportunities for people to express these social aspects of individual deliberation. After all, freedom of assembly, speech, and association are fundamental liberal rights. The opportunities for collective enquiry simply occur within and between groups and associations below the level of the state—friends and family, in the first instance, but also churches, cultural associations, professional groups and trade unions, universities, and the mass media. Liberals do not deny that 'the public display of character and judgement and the exchange of experience and insight' are needed to make intelligent judgements about the good, or to show others that I 'hold [my] notion of the good responsibly' (Crowley 1987: 285). Indeed, these claims fit comfortably in many liberal discussions of the value of free speech and association (e.g. Scanlon 1983: 141–7). What the liberal denies is that I should have to give such an account of myself *to the state*.

A similar failure to consider the distinctive role of the state weakens radical critiques of liberal neutrality, like that of Habermas. Habermas wants the evaluation of different ways of life to be a political question, but unlike communitarians, he does not hope that this political deliberation will serve to promote people's embeddedness in existing practices. Indeed, he thinks that political deliberation is required precisely because in its absence people will tend to accept existing practices as givens, and thereby perpetuate the false needs and false consciousness which accompany those historical practices.[8] Only when existing ways of life are 'the objects of discursive will-formation' can people's understanding of the good be free of deception. Neutrality does not demand the scrutiny of these practices, and hence does not recognize the emancipatory interest people have in escaping false needs and ideological distortions.

But why should the evaluation of people's conceptions of the good be tied to their claims of justice and hence to the state? Communities smaller than the entire political society, groups and associations of various sizes, might be more appropriate forums for those forms of 'discursive will-formation' which involve evaluating the good, and interpreting one's genuine needs. While Habermas rejects the communitarian tendency to endorse existing social practices uncritically as the basis for political deliberations about the good, he shares their tendency to assume that anything which is not politically deliberated is thereby left to an individual will incapable of rational judgement.

So liberal neutrality does not neglect the importance of a shared culture for meaningful individual options, or of the sharing of experiences for meaningful individual evaluation of those options. Liberal neutrality does not deny these social requirements of individual autonomy, but rather provides an interpretation of them, one which relies on social rather than political processes. None of this proves that neutrality should be endorsed. Neutrality requires a certain faith in the operation of non-state forums and processes for individual judgement and cultural development, and a distrust of the operation of state forums for evaluating the good. Nothing I have said shows that this optimism and distrust are warranted. Indeed, just as critics of neutrality have failed to defend their faith in politics, so liberals have failed to defend their faith in non-state forums.

In fact, it seems that each side in the neutrality debate has failed to learn the lesson taught by the other side. Despite centuries of liberal insistence on the importance of the distinction between state and society, communitarians still seem to assume that whatever is properly social must become the province of the political. They have not confronted the liberal worry that the all-embracing authority and coercive means which characterize the state make it a particularly inappropriate forum for the sort of genuinely shared deliberation and commitment that they desire. Despite centuries of communitarian insistence on the historically fragile nature of our culture, and the need to consider the conditions under which a free culture can sustain itself, liberals still tend to take the existence of a tolerant and diverse culture for granted, as something which naturally arises and sustains itself, the ongoing existence of which is therefore simply assumed in a theory of justice. Communitarians are right to insist that a culture of freedom is a historical achievement, and liberals need to explain why the cultural marketplace does not threaten that achievement either by failing to connect people in a strong enough way to their communal practices (as communitarians fear), or conversely, by failing to detach people in a strong enough way from the expectations of existing practices and ideologies (as Habermas fears). A culture which supports self-determination requires a mix of both exposure and connection to existing practices, and also distance and dissent from them. Liberal neutrality may provide that mix, but that is not obviously true, and it may be true only in some times and places. So both sides need to give us a more comprehensive comparison of the opportunities and dangers present in state and non-state forums and procedures for evaluating the good.

I have argued elsewhere that before invoking the state as the arena for evaluating conceptions of the good, we should first improve the forums in civil society for non-politicized debate, so as to ensure that all groups in society have genuinely free and equal access to the cultural marketplace that liberals value so highly (Kymlicka 1989*b*). But while this question remains open, it should be clear that we are not likely to get anywhere in answering it if we continue to see it as a debate between liberal 'atomism' and communitarianism's 'social thesis'. According to communitarians, liberals fail to recognize that people are naturally social beings. Liberals supposedly think that society rests on an artificial social contract, and that state power is needed to keep naturally asocial people together in society. But there is a sense in which the opposite is true—liberals believe that people naturally form and join social relations and forums in which they come to understand and pursue the good. The state is not needed to provide that communal context, and is likely to distort the normal processes

of collective deliberations and cultural development. It is communitarians who seem to think that individuals will drift into
anomic isolation without the state actively bringing them together
to evaluate and pursue the good.[9]

(c) Political legitimacy

There is another issue raised by the social thesis. Individual choices
require a secure cultural context, but a cultural context, in turn,
requires a secure political context. Whatever the proper role of the
state in protecting the cultural marketplace, it can only fulfil that
function if public institutions are stable, and that in turn requires
that they have legitimacy in the eyes of the citizens. Taylor believes
that political institutions governed by the principle of neutrality
will be incapable of sustaining legitimacy, and hence incapable of
sustaining the social context required for self-determination.

According to Taylor, the neutral state undermines the shared
sense of the common good which is required for citizens to accept
the sacrifices demanded by the welfare state. Citizens will only
identify with the state, and accept its demands as legitimate, when
there is a 'common form of life' which 'is seen as a supremely
important good, so that its continuance and flourishing matters to
the citizens for its own sake and not just instrumentally to their
several individual goods or as the sum total of these individual
goods' (Taylor 1986: 213). But this sense of the common good has
been undermined because, in part, we now have a political culture
of state neutrality in which people are free to choose their goals
independently of this 'common form of life', and to trump the
pursuit of this common good should it violate their rights. Whereas
a communitarian state would foster an identification with the
common form of life, the

> rights model goes very well with a more atomist consciousness, where I
> understand my dignity as that of an individual bearer of rights. Indeed—
> and here the tension surfaces between the two—I cannot be too willing to
> trump the collective decision in the name of individual rights if I haven't
> already moved some distance from the community which makes these
> decisions. (Taylor 1986: 211)

This 'distancing' from the community's shared form of life means
we become unwilling to shoulder the burdens of liberal justice. As a

result, liberal democracies are undergoing a 'legitimation crisis'—citizens are asked to sacrifice more and more in the name of justice, but they share less and less with those for whom they are making sacrifices. There is no shared form of life underlying the demands of the neutral state.

Rawls and Dworkin, on the other hand, believe that citizens will accept the burdens of justice even in their relations with people who have very different conceptions of the good. A person should be free to choose any conception of the good life that does not violate the principles of justice, no matter how much it differs from other ways of life in the community. Such conflicting conceptions can be tolerated because the public recognition of principles of justice is sufficient to ensure stability even in the face of such conflicts (Rawls 1985: 245). People with different conceptions of the good will respect each other's rights, not because it promotes a shared way of life, but because citizens recognize that each person has an equal claim to consideration. Hence the basis for state legitimacy is a shared sense of justice, not a shared conception of the good. Liberals seek to sustain a just society through the public adoption of principles of justice, without requiring, and indeed precluding, the public adoption of certain principles of the good life.

Taylor believes this is sociologically naïve: people will not respect the claims of others unless they are bound by shared conceptions of the good, unless they can identify with a politics of the common good. He describes 'two package solutions emerging out of the mists to the problem of sustaining a viable modern polity in the late twentieth century', which correspond roughly to the communitarian and liberal model, and says there are 'severe doubts' about the long-term viability of the liberal model. By enforcing individual rights and state neutrality, a liberal state precludes the public adoption of principles of the good, but, Taylor asks, 'Could the increasing stress on rights as dominant over collective decisions come in the end to undermine the very legitimacy of the democratic order?' (Taylor 1986: 225).

Why is a shared way of life required to sustain legitimacy? Taylor does not give any clear-cut explanation of the need for a specifically communitarian politics.[10] But one answer that is implicit in communitarian writings lies in a romanticized view of earlier societies in which legitimacy was based on the effective pursuit of shared ends. Communitarians imply that we could recover the sense of allegiance that was present in earlier days if we accepted a politics of the common good, and encouraged everyone to participate freely in it. Common examples of such earlier societies are the republican democracies of Ancient Greece, or eighteenth-century New England town governments.

But these historical examples ignore an important fact. Early New England town governments may have had a great deal of legitimacy amongst their members in virtue of the effective pursuit of their shared ends. But that is at least partly because women, atheists, Indians, and the propertyless were all excluded from membership. Had they been allowed membership, they would not have been impressed by the pursuit of what was often a racist and sexist 'common good'. The way in which legitimacy was ensured amongst all members was to exclude some from membership.

Contemporary communitarians are not advocating that legitimacy be secured by denying membership to those groups in the community who have not historically participated in shaping the 'common way of life'. Communitarians believe that there are certain communal practices that everyone can endorse as the basis for a politics of the common good. But what are these practices? Communitarians often write as if the historical exclusion of certain groups from various social practices was just arbitrary, so that we can now include them and proceed forward. But the exclusion of women, for example, was not arbitrary. It was done for a reason—namely, that the ends being pursued were sexist, defined by men to serve their interests. Demanding that women accept an identity that men have defined for them is not a promising way to increase their sense of allegiance. We cannot avoid this problem by saying with Sandel that women's identities are constituted by existing roles. That is simply false: women can and have rejected those roles, which in many ways operate to deny their separate identity. That was also true in eighteenth-century New England, but legitimacy there was preserved by excluding women from membership. We must find some other way of securing legitimacy, one that does not continue to define excluded groups in terms of an identity that others created for them.

Sandel and Taylor say that there are shared ends that can serve as the basis for a politics of the common good which will be legitimate for all groups in society. But they give no examples of such ends— and surely part of the reason is that there are none. They say that these shared ends are to be found in our historical practices, but they do not mention that those practices were defined by a small section of society—propertied white men—to serve the interests of propertied white men. These practices are gender-coded, race-coded, and class-coded, even when women, blacks, and workers are legally allowed to participate in them. Attempts to promote these kinds of ends reduce legitimacy, and further exclude marginalized groups. Indeed, just such a loss of legitimacy seems to be occurring amongst many elements of American society—blacks, gays, single mothers, non-Christians—as the right wing tries to implement its agenda based on the Christian, patriarchal family. Many communitarians undoubtedly dislike the Moral Majority's view of the common good, but the problem of the exclusion of historically marginalized groups is endemic to the communitarian project. As Hirsch notes, 'any "renewal" or strengthening of community sentiment will accomplish nothing for these groups'. On the contrary, our historical sentiments and traditions are 'part of the problem, not part of the solution' (Hirsch 1986: 424).

Consider one of the few concrete examples of communitarian politics that Sandel offers—the regulation of pornography. Sandel argues that such regulation by a local community is permissible 'on the grounds that pornography offends its way of life' (Sandel 1984*b*: 17). To consider how exclusionary this argument can be, contrast it with recent feminist discussions of pornography. Many women's groups have demanded the regulation of pornography on the grounds that women have been excluded from the process of defining traditional views of sexuality. Pornography, some feminists argue, plays a critical role in promoting violence against women, and in perpetuating the subordination of women to male-defined ideas of sexuality and gender roles (e.g. Mackinnon 1987: ch. 13–14). This argument is controversial, but if pornography does in fact play this role in the subordination of women, it does so not because it 'offends our way of life', but precisely because it conforms to our cultural stereotypes about sexuality and the role of women. In fact, as Mackinnon notes, from a feminist point of view the problem with pornography is not that it violates community standards but that it enforces them.

Sandel's argument is in fundamental conflict with this feminist argument. The problem with Sandel's view can be seen by considering the regulation of homosexuality. Homosexuality is 'offensive to the way of life' of many Americans. Indeed, measured by any plausible standard, more people are offended by homosexuality than by pornography. Would Sandel therefore allow local communities to criminalize homosexual relations, or the public affirmation of homosexuality? If not, what distinguishes it from pornography? For liberals, the difference is that homosexuality does not harm others, and the fact that others are offended by it has no moral weight. The majority in a local (or national) community does not have the right to enforce its external preferences concerning the practices of those people who are outside the mainstream way of life (Dworkin 1985: 353–72; cf. ch. 2, s. 5a above). But this is precisely what Sandel cannot say. On his argument, members of marginalized groups must adjust their personalities and practices so as to be inoffensive to the dominant values of the community. Nothing in Sandel's argument gives members of marginalized groups the power to reject the identity that others have historically defined for them.[11]

Likewise, in the case of pornography, Sandel is not affirming the importance of giving women the ability to reject the male view of sexuality, and to define their own sexuality. On the contrary, he is saying that pornography can be regulated whenever one male-defined view of sexuality (the pornographers') conflicts with another male-defined view of sexuality (the 'way of life' of the community). And nothing guarantees that the men who are offended by pornography will not have a different but equally oppressive view of female sexuality (e.g. the fundamentalist view that women's sexuality must be kept strictly repressed). However the community decides, women, like all marginalized groups, will have to adjust their aims to be inoffensive to a way of life that they had little or no role in defining. This is no way to develop feelings of legitimacy amongst members of marginalized groups.

Communitarians like to say that political theory should pay more attention to the history of each culture. But it is remarkable how rarely communitarians themselves undertake such an examination of our culture. They wish to use the ends and practices of our cultural tradition as the basis for a politics of the common good, but they do not mention that these practices were defined by a small segment of the population. If we look at the history of our society, surely liberal neutrality has the great advantage of its potential inclusiveness, its denial that subordinated groups must fit into the 'way of life' that has been defined by the dominant groups. Communitarians simply ignore this danger and the history which makes it so difficult to avoid.[12]

Sandel concludes his book by saying that when politics goes well 'we can know a good in common that we cannot know alone' (Sandel 1982: 183). But given the diversity of modern societies, we should say instead that politics goes well precisely when it does not adopt an ideology of the 'common good' that can only serve to exclude many groups. Increasing the level of state legitimacy may well require greater civic participation by all groups in society, but, as Dworkin notes, it only makes sense to invite people to participate in politics (or for people to accept that invitation) if they will be treated as equals (Dworkin 1983: 33). And that is incompatible with defining people in terms of roles they did not shape or endorse. If legitimacy is to be earned, it will not be by strengthening communal practices that have been defined by and for others. It will require empowering the oppressed to define their own aims. Liberalism may not do enough in this regard, but as Herzog puts it, if liberalism is the problem, how could communitarianism be the solution? (Herzog 1986: 484).

Whether liberal politics would in fact sustain a sense of public legitimacy is difficult to determine, since liberal principles remain to be properly implemented. But I believe that liberal neutrality is the most likely principle to secure public assent in societies like ours, which are diverse and historically exclusionary. Inviting people to participate in politics on any other understanding is not likely to be successful. As Mill said, a feeling of commitment to a common public philosophy is a precondition of a free culture, and 'the only shape in which the feeling is likely to exist hereafter' is an attachment to 'the principles of individual freedom and political and social equality, as realized in institutions which as yet exist nowhere, or exist only in a rudimentary state' (Mill 1962: 122–3). Those principles remain largely unrealized in practice, but they are, more than ever, the only viable basis for public legitimacy.[13]

Communitarians are right to emphasize the importance of the social thesis, and hence the importance of a secure social context for the exercise of our capacities for choice. And they are right to claim that this in turn involves the need for civic participation and political legitimacy. All these are of unquestionable importance. But that is just the problem. No one does question their importance. Liberals and communitarians disagree, not over the social thesis, but over the proper role of the state. Is the role of the state to protect 'the free internal life of the various communities of interests in which persons and groups seek to achieve . . . the ends and excellences to which they are drawn' (Rawls 1971: 543), or should it also partly pre-empt that social life, by enforcing a public ranking of the ends and excellences to which they should be drawn? To (over)simplify, liberals and communitarians disagree, not over the individual's dependence on society, but over society's dependence on the state. This is an important debate, but it is not a debate between those who do and those who do not accept the social thesis. In fact, it is a number of different debates, each of which should be considered on its own, and each of which requires more empirical argument than either side has felt disposed to provide.[14]

5. THE POLITICS OF COMMUNITARIANISM

The communitarian theory of a politics of the common good arises out of some important practical concerns. While liberal theory may recognize the dependence of individual choice on the cultural context, liberals have in practice focused their concern on individual freedom of choice to the neglect of people's access to culture. Liberal organizations, such as the American Civil Liberties Union, have been preoccupied with threats to the right of free speech, like restrictions on hate literature or obscene material. But surely the fact that 10 per cent of adults are functionally illiterate is a more serious threat to people's unfettered participation in the cultural marketplace than the restrictions on obscenity. And the fact that ownership of the media is so concentrated that a number of viewpoints get systematically silenced is a more serious threat to the free exchange of information than restrictions on hate literature. Liberals have often operated with confused priorities in the area of culture.

Given these failures to ensure that everyone has meaningful access to the cultural achievements and collective deliberations of the community, the communitarian desire to create a language and practice of a politics of the common good is understandable. Unfortunately, such a language and practice is at best irrelevant to modern democracies, and intolerant at worst. In fact, both liberals and communitarians have ignored the real issues involved in creating the cultural conditions for self-determination.

Consider the question of language. Both communitarians and liberals operate, implicitly or explicitly, with the assumption that all states are 'nation-states'—that everyone in each country shares the same nationality, and so speaks the same language, and can join in a meaningful debate about culture. But most countries are multi-nation states, containing two or more linguistic communities. In Canada, for example, there are French and English, as well as the aboriginal languages retained by Indian and Inuit communities. Should we attempt to create linguistic homogeneity, in the interests of either liberal equality or the communitarian common good? Some liberals and communitarians have assumed that this is the goal, and so have defended programmes for assimilating linguistic minorities. But others have opposed assimilation as patently unfair (why should the French, or native Indians, have to assimilate to English, especially when they were in Canada before the English?). But if we allow minority cultures to exist, what rights do they have with respect to their language? Both the French and native Indian peoples of Canada have special legal rights designed to protect their distinct cultures—for example, the right to public education in their own language, and the right to use their language in dealing with government and the courts. Moreover, they have the power to impose restrictions on the language rights of non-Francophones or non-aboriginals who move into their homelands. Special rights and/or separate institutions also exist for minority language groups in the United States (e.g. the Puerto Ricans, the American Indians), for the aboriginal populations in Australia and New Zealand, and for minority language groups in Belgium, Switzerland, and most Second and Third World countries.

In all of these countries, the question of whose language is used by the state in schools, courts, and bureaucracies is an important and divisive one. Indeed, this question has been the primary source of conflict in many of these countries. Yet one searches in vain amongst contemporary liberals and communitarians to find a single discussion of this question. They debate what role the state should play in promoting 'its culture' and enriching 'its language' (e.g. Dworkin 1985: 230–3; Sullivan 1982: 173), but they never ask *whose* culture and *which* language. They debate whether the schools should promote particular conceptions of the good, but they do not ask what language the schools should be using. If they began to ask these basic questions, much of what passes for the received wisdom concerning the relationship between state and culture would quickly be rendered obsolete. Indeed, I have argued elsewhere that much of the received wisdom concerning the meaning of equal rights and anti-discrimination must also be abandoned in multi-national states.[15] The fact is that we do not know what either liberal neutrality or the communitarian common good requires in multi-nation states. This is perhaps the most glaring example of how the communitarians' emphasis on the social thesis has become detached from any actual examination of the connections between the individual, culture, and the state.

NOTES

1. Certain acts of paternalism involving competent adults may be justified when we are faced with clear cases of weakness of will. For example, most people know that the gain in safety is well worth the effort of putting on a car seat-belt. Yet many people let momentary inconvenience override their reason. Mandatory seat-belt legislation helps overcome this weakness of will, by giving people an extra incentive to do something that they know they already have sufficient reason to do.

2. The case of coerced religious worship has been a favourite example of liberals from Locke to Rawls. It is not clear that religious worship can be generalized in this way, since there is an epistemic requirement to praying that is not always present elsewhere. However, I believe that the 'endorsement constraint' is applicable to most valuable and important forms of human activity (Dworkin 1989: 484–7; Raz 1986: 291–3, but cf. Daniels 1975: 266). Some liberals argue that the endorsement constraint makes perfectionism necessarily self-defeating. For even if the state can encourage or force people to pursue the most valuable ways of life, it cannot get people to pursue them for the right reasons. Someone who changes their lifestyle in order to avoid state punishment, or to gain state subsidies, is not guided by an understanding of the genuine value of the new activity (Waldron 1989: 1145–6; Lomasky 1987: 253–4). This is a valid point against coercive and manipulative forms of perfectionism. But it does not rule out short-term state intervention designed to introduce people to

valuable ways of life. One way to get people to pursue something for the right reasons is to get them to pursue it for the wrong reasons, and hope they will then see its true value. This is not inherently unacceptable, and it occurs often enough in the cultural marketplace. So the endorsement constraint argument, by itself, cannot rule out all forms of state perfectionism.

3. The importance of the revisability of our ends has played a changing role in Rawls's theory. In his most recent articles, Rawls has qualified his original view. He now says that we should accept that view for the purposes of determining our public rights and responsibilities, without necessarily accepting it as an accurate portrayal of our private self-understandings (Rawls 1980: 545; 1985: 240–4; cf. Buchanan 1982: 138–44). I will concentrate on Rawls's original presentation, partly because it is the one to which most communitarians have responded, and partly because I believe that his more recent view is unsuccessful (Kymlicka 1989a: 58–61).

4. For other major statements of liberal neutrality, see Ackerman (1980: 11, 61); Larmore (1987: 44–7); Dworkin (1978: 127; 1985: 222). 'Neutrality' may not be the best word to describe the policy at issue. Rawls himself has avoided the term because of its multiple and often misleading meanings. Neutrality in everyday usage often refers to the consequences of actions, rather than the justifications for them (Rawls 1988: 260, 265; cf. Raz 1986: ch. 5). A 'neutral' policy, in this everyday sense, would be one that ensured that all conceptions of the good fared equally well in society, no matter how expensive and unattractive they were. Some critics have taken Rawls to defend neutrality in this everyday sense (e.g. Raz 1986: 117). However, this sort of neutrality is quite illiberal, since it would both restrict freedom of choice, and violate the requirement that people accept responsibility for the costs of their choices. Any society which allows different ways of life to compete for people's free allegiance, and which requires people to pay for the costs of their choices, will seriously disadvantage expensive and unattractive ways of life. Liberals accept, and indeed value, these unequal consequences of civil liberties and individual responsibility. Hence liberal neutrality—i.e. the requirement that the liberal state do not rank the value of different ways of life—is different from, and indeed opposed to, neutrality in the everyday sense (cf. Kymlicka 1989b: 883–6). To avoid this possible misinterpretation, Rawls has instead used the term 'priority of the right over the good'. But that too has multiple and misleading meanings, since it is used by Rawls to describe both the affirming of neutrality over perfectionism, and the affirming of deontology over teleology. These issues need to be kept distinct, and neither, viewed on its own, is usefully called a matter of the 'priority of the right'. See Kymlicka (1988b: 173–90) for a

critique of Rawls's usage of 'priority of the right'. Given the absence of any obviously superior alternative, I will continue to use the term 'neutrality'.

5. There is another objection that deserves mention, concerning the need for social confirmation of our individual judgements. According to some communitarians, while it may be important for individuals to endorse the value of their activities from the inside, it is equally if not more important that other people confirm that judgement from the outside. Without outside confirmation, we lose our sense of self-respect, our confidence in the value of our own judgements. A communitarian state, therefore, would limit Dworkin's 'individual endorsement' constraint where it unduly threatens the communitarian 'social confirmation' constraint. I discuss this in Kymlicka (1988*a*: 195–7); cf. Williams (1985: 169–70); Smith (1985: 188–92); Dworkin (1987: 16–17).

6. I have focused on Sandel's writings, but the same ambiguity in the communitarian theory of the self can also be found in MacIntyre (1981: 200–6) and Taylor (1979: 157–60). See Kymlicka (1989*a*: 56–7) for a discussion of these writers. Sandel's claim that Rawls's view of the self violates our self-understanding gets much of its force from being linked to the further claim that Rawls views people as being essentially disembodied. According to Sandel, the reason that Rawls denies that people are entitled to the rewards which accrue from the exercise of their natural talents is that he denies that natural talents are an essential part of our personal identity. They are mere possessions, not constituents, of the self (Sandel 1982: 72–94; Larmore 1987: 127). But this is a misinterpretation. The reason why Rawls denies that people are entitled to the fruits of the exercise of their natural talents is that no one deserves their place in the natural lottery, no one deserves greater natural talents than anyone else (ch. 3, s. 2 above). This position is entirely consistent with the claim that natural talents are constituents of the self. The fact that natural talents are constitutive of the self does nothing to show that a gifted child deserved to be born more talented than an ordinary child. Many liberals would not accept the claim that all our natural attributes are constituents of the self (e.g. Dworkin 1983: 39), and I myself am unsure where to draw this line (ch. 4, s. 5 above). But wherever we draw this line, the ways in which we are essentially *physically embodied* does nothing to support Sandel's conception of the ways that we are *socially embedded*.

7. Taylor says that he is criticizing the 'primacy of rights' doctrine, by which he means the claim that individual rights have primacy over other moral notions, such as individual duties, the common good, virtue, etc. According to Taylor, this doctrine is found in Hobbes, Locke, and Nozick. I find this schema unhelpful, for none of these

moral notions, including individual rights, is the right sort to be morally primary. (Notice, for example, that Hobbes and Nozick are both 'primacy of rights' theorists on Taylor's schema. But since Nozick affirms what Hobbes denies—i.e. that individuals have inherent moral standing—any agreement between them over individual rights must be derivative, not morally primary: cf. ch. 4, s. 3 above.) The debate that Taylor wishes to consider is best pursued by asking not whether rights in general have primacy over duties in general, but whether there are particular rights, duties, virtues, etc. that are inadequately recognized in liberal or (as Taylor describes them) 'ultra-liberal' theories. And if we look at the debate this way, one of Taylor's arguments is that state neutrality can undermine the social conditions necessary for individual autonomy. That claim, if true, has importance for liberal and libertarian theories whether or not they endorse the 'primacy of rights' doctrine.

Some communitarians take the social thesis to undermine liberalism in a more fundamental way, by undermining its moral individualism. Moral individualism is the view that individuals are the basic unit of moral value, so that any moral duties to larger units (e.g. the community) must be derived from our obligations to individuals. But, communitarians argue, if we reject the atomistic view that individuals are self-originating persons, then we must also reject Rawls's claim that we are 'self-originating sources of valid claims' (Rawls 1980: 543). But this is a *non sequitur*. Rawls's claim that we are self-originating sources of valid claims is not a sociological claim about how we develop. It is a moral claim about the location of moral value. As Galston says, 'while the formative power of society is surely decisive, it is nevertheless *individuals* that are being shaped. I may share everything with others. But it is *I* that shares them—an independent consciousness, a separate locus of pleasure and pain, a demarcated being with interests to be advanced or suppressed' (Galston 1986: 91). While my good is socially determined, it is still my good that is affected by social life, and any plausible political theory must attend equally to the interests of each person.

8. Habermas seems to endorse this position when he says that the need for a 'discursive desolidification of the (largely externally controlled or traditionally fixed) interpretations of our needs' is the heart of his disagreement with Rawls (Habermas 1979: 198–9). However, he now rejects the idea of politically evaluating people's conceptions of the good (Habermas 1985: 214–16; cf. Benhabib 1986: 332–43; Funk 1988: 29–31).

9. The suggestion that non-political activity is inherently solitary is implicit in a number of communitarian writings. For example, Sandel claims that under communitarian politics 'we can know a good in

common that we cannot know alone' (Sandel 1982: 183). And Sullivan claims that state perfectionism is needed to ensure that no one is 'cut off' from collective deliberations (Sullivan 1982: 158). Liberals make the opposite assumption that the state is not required to lead individuals into collective associations and deliberations (Macedo 1988: 127–8; Feinberg 1988: 105–13).

10. Taylor does suggest some preconditions of legitimacy which he believes are not attended to in liberal neutrality—in particular, the need for citizen participation (e.g. Taylor 1986: 225; 1989: 177–81). But he does not adequately explain why liberal neutrality cannot serve this requirement. His real target, I think, is a certain kind of 'bureaucratic tutelage' which subordinates democratic politics to the rule of experts (Taylor 1989: 180). But that problem is not a distinctively liberal one.

11. In a recent article, Sandel suggests that American laws against sodomy should be overturned on the ground that some homosexual relations aim at the same substantive ends as characterize heterosexual marriages, which have traditionally received Supreme Court protection (Sandel 1989: 344–5). But why should the freedom of homosexuals depend on their pursuing the same aims and aspirations as heterosexuals? Many gay rights groups would deny that they have the same (restrictive) view of intimacy and sexuality as that which characterizes traditional heterosexual marriages. What if, as the Supreme Court argued in a recent case upholding anti-sodomy laws, gay rights threaten the perceived sanctity of the heterosexual family? In any event, Sandel does not explain how his new argument that anti-sodomy laws are unconstitutional fits in with his earlier claim about the freedom of local communities to regulate activities that offend their way of life.

12. On the exclusionary tendencies of communitarianism, see Gutmann (1985: 318–22); Herzog (1986: 481–90); Hirsch (1986: 435–8); Rosenblum (1987: 178–81). I have argued elsewhere that many of these considerations also argue against non-communitarian forms of perfectionist intervention in the cultural marketplace. Even where it is not deliberately aimed at promoting the community's way of life, state perfectionism would tend to distort the free evaluation of ways of life, to rigidify the dominant ways of life, regardless of their intrinsic merits, and to exclude unfairly the values and aspirations of marginalized and disadvantaged groups within our society (Kymlicka 1989b: 900–2).

13. Rawls cites the need for public legitimacy as grounds for supporting rather than opposing neutrality. He claims that perfectionism threatens the public consensus, because people will not accept the legitimacy of state policies based on a conception of the good they do not share.

Rawls seems to think that this will be true of any society where citizens are divided by conflicting conceptions of the good. Put at this general level, Rawls's claim is surely false. As Raz shows, it is possible for people with conflicting ends to agree none the less on a procedure for arriving at a public ranking of the value of different ways of life, or perhaps to accept a particular public ranking with which they disagree but which they none the less see as a better second-best option than neutrality (Raz 1986: 126–32). There is no inherent connection between neutrality and state legitimacy. However, the kinds of conflicting ends in modern democracies, and the history underlying them, are such that perfectionism of the communitarian variety surely is a threat to state legitimacy.

14. For helpful attempts to break down the debate into different empirical issues, see Buchanan (1989) and Walzer (1990). For a philosophically informed attempt to provide empirical support for the communitarian position, see Bellah *et al.* (1985); but cf. Macedo (1988); Stout (1986).

15. The assumption that the political community is culturally homogeneous is present in a number of passages in Rawls and Dworkin (Rawls 1978: 55; Dworkin 1985: 230–3). While revising that assumption would affect the conclusions they go on to draw about people's rights, Rawls and Dworkin never discuss what changes would be required in culturally plural countries. Indeed, they do not seem to recognize that any changes would be required. I provide a liberal theory of the rights of minority cultures in Kymlicka (1989a: chs. 7–10); cf. Van Dyke (1975).

7

Feminism

Contemporary feminist political theory is extremely diverse, in both premises and conclusions. This is also true to some degree of the other theories I have examined. But this diversity is multiplied within feminism, for each of these other theories is represented within feminism. Thus we have liberal feminism, socialist feminism, even libertarian feminism. Moreover, there is a significant movement within feminism towards forms of theorizing, such as psychoanalytic or post-structuralist theory, which lie outside the bounds of mainstream Anglo-American political philosophy. Alison Jaggar says that a commitment to eliminating the subordination of women unifies the diverse strands of feminist theory (Jaggar 1983: 5). But (as Jaggar notes) this agreement soon dissolves into radically different accounts of that subordination, and of the measures required to eliminate it.

It would require a separate book to discuss each of these strands of feminist theory.[1] I will instead focus on three feminist criticisms of the way mainstream political theories attend, or fail to attend, to the interests and concerns of women. I have argued that a wide range of contemporary political theories share an 'egalitarian plateau', a commitment to the idea that all members of the community should be treated as equals. Yet, until very recently, most mainstream political philosophy has defended, or at least accepted, sexual discrimination. And while traditional views about sexual discrimination have been progressively abandoned, many feminists believe that the principles which were developed with men's experience and interests in mind are incapable of adequately recognizing women's needs, or incorporating women's experiences. I will consider three such arguments. The first focuses on the 'gender-neutral' account of sexual discrimination; the second focuses on the public–private distinction. These two arguments claim that important aspects of the liberal-democratic conception of justice are male-biased. The third argument, on the other hand, claims that the very emphasis on justice is itself reflective of a male-bias, and that any theory which is responsive to the interests and experiences of women will replace the emphasis on justice with an emphasis on caring. These three arguments give only a limited idea of the scope of recent feminist theory, but they raise important issues which any account of sexual equality must address, and they represent three of the most sustained points of contact between feminism and mainstream political philosophy.

1. SEXUAL EQUALITY AND DISCRIMINATION

Until well into this century, most male theorists on all points of the political spectrum accepted the belief that there was a 'foundation in nature' for the confinement of women to the family, and for the 'legal and customary subjection of women to their husbands' within the family (Okin 1979: 200).[2] Restrictions on women's civil and political rights were said to be justified by the fact that women are, by nature, unsuited for political and economic activities outside the home. Contemporary theorists have progressively abandoned this assumption of women's natural inferiority. They have accepted that women, like men, should be viewed as 'free and equal beings', capable of self-determination and a sense of justice, and hence free to enter the public realm. And liberal democracies have progressively adopted anti-discrimination statutes intended to ensure that women have equal access to education, employment, political office, etc.

But these anti-discrimination statutes have not brought about sexual equality. In the United States and Canada, the extent of job segregation in the lowest-paying occupations is increasing. Indeed, if present trends continue, all of the people below the poverty line in America in the year 2000 will be women or children (Weitzman 1985: 350). Moreover, domestic violence and sexual assault are increasing, as are other forms of violence and degradation aimed at women. Catherine Mackinnon summarizes her survey of the effects of equal rights in the United States by saying that 'sex equality law has been utterly ineffective at getting women what we need and are socially prevented from having on the basis of a condition of birth:

a chance at productive lives of reasonable physical security, self-
expression, individuation, and minimal respect and dignity'
(Mackinnon 1987: 32).

Why is this? Sex discrimination, as commonly interpreted,
involves the arbitrary or irrational use of gender in the awarding of
benefits or positions. On this view, the most blatant forms of sex
discrimination are those where, for example, someone refuses to
hire a woman for a job even though gender has no rational
relationship to the task being performed. Mackinnon calls this the
'difference approach' to sexual discrimination, for it views as
discriminatory unequal treatment that cannot be justified by
reference to some sexual difference.

Sex discrimination law of this sort was modelled on race
discrimination law. And just as race equality legislation aims at a
'colour-blind' society, so sex equality law aims at a sex-blind
society. A society would be non-discriminatory if race or gender
never entered into the awarding of benefits. Of course, while it is
conceivable that political and economic decisions could entirely
disregard race, it is difficult to see how a society could be entirely
sex-blind. A society which provides for pregnancy benefits, or for
sexually segregated sports, is taking sex into account, but this does
not seem unjust. And while racially segregated washrooms are
clearly discriminatory, most people do not feel that way about sex-
segregated washrooms. So the 'difference approach' accepts that
there are legitimate instances of differential treatment of the sexes.
These are not discriminatory, however, so long as there is a genuine
sexual difference which explains and justifies the differential
treatment. Opponents of equal rights for women often invoked the
spectre of sexually integrated sports (or washrooms) as evidence
that sex equality is misguided. But defenders of the difference
approach respond that the cases of legitimate differentiation are
sufficiently rare, and the cases of arbitrary differentation so
common, that the burden of proof rests on those who claim that sex
is a relevant ground for assigning benefits or positions.

This difference approach, as the standard interpretation of sex
equality law in most Western countries, has had some successes. Its
'moral thrust' is to 'grant women access to what men have access
to', and it has indeed 'gotten women some access to employment
and education, the public pursuits, including academic, professional, and blue-collar work, the military, and more than nominal

access to athletics' (Mackinnon 1987: 33, 35). The difference approach has helped create gender-neutral access to, or competition for, existing social benefits and positions.

But its successes are limited, for it ignores the gender inequalities which are built into the very definition of these positions. The difference approach sees sex equality in terms of the ability of women to compete under gender-neutral rules for the roles that men have defined. But equality cannot be achieved by allowing men to build social institutions according to their interests, and then ignoring the gender of the candidates when deciding who fills the roles in these institutions. The problem is that the roles may be defined in such a way as to make men more suited to them, even under gender-neutral competition.

Consider the fact that most jobs 'require that the person, gender neutral, who is qualified for them will be someone who is not the primary caretaker of a preschool child' (Mackinnon 1987: 37). Given that women are still expected to take care of children in our society, men will tend to do better than women in competing for such jobs. This is not because women applicants are discriminated against. Employers may pay no attention to the gender of the applicants, or may in fact wish to hire more women. The problem is that many women lack a relevant qualification for the job—i.e. being free from child-care responsibilities. There is gender-neutrality, in that employers do not attend to the gender of applicants, but there is no sexual equality, for the job was defined under the assumption that it would be filled by men who had wives at home taking care of the children. The difference approach insists that gender should not be taken into account in deciding who should have a job, but it ignores the fact 'that day one of taking gender into account was the day the job was structured with the expectation that its occupant would have no child care responsibilities' (Mackinnon 1987: 37).

Whether or not gender-neutrality yields sexual equality depends on whether and how gender was taken into account earlier. As Janet Radcliffe Richards says,

If a group is kept out of something for long enough, it is overwhelmingly likely that activities of that sort will develop in a way unsuited to the excluded group. We know for certain that women have been kept out of many kinds of work, and this means that the work is quite likely to be unsuited to them. The most obvious example of this is the incompatibility of most work with the bearing and raising of children; I am firmly convinced that if women had been fully involved in the running of society from the start they would have *found* a way of arranging work and children to fit each other. Men have had no such motivation, and we can see the results. (Radcliffe Richards 1980: 113–14)

This incompatibility that men have created between child-rearing and paid labour has profoundly unequal results for women. The result is not only that the most valued positions in society are filled by men, while women are disproportionately concentrated into lower-paying part-time work, but also that many women become economically dependent on men. Where most of the 'household income' comes from the man's paid work, the woman who does the unpaid domestic work is rendered dependent on him for access to resources. The consequences of this dependence have become more apparent with the rising divorce rate. While married couples may share the same standard of living during marriage, regardless of who earns the income, the effects of divorce are catastrophically unequal. In California, men's average standard of living goes up 42 per cent after divorce, women's goes down 73 per cent, and similar results have been found in other states (Okin 1989*b*: 161). However, none of these unequal consequences of the incompatibility of child care and paid work are discriminatory, according to the difference approach, for they do not involve arbitrary discrimination. The fact is that freedom from child-care responsibilities is relevant to most existing jobs, and employers are not being arbitrary in insisting on it. Because it is a relevant qualification, the difference approach says that it is not discriminatory to insist upon it, regardless of the disadvantages it creates for women. Indeed, the difference approach sees the concern with child-care responsibilities, rather than irrelevant criteria like gender, as evidence that sex discrimination has been eliminated. It cannot see that the relevance of child-care responsibilities is itself a profound source of sexual inequality, one that has arisen from the way men have historically structured the economy to suit their interests.

So before we decide whether gender should be taken into account, we need to know how it has already been taken into account. And the fact is that almost all important roles and positions have been structured in gender-biased ways:

virtually every quality that distinguishes men from women is already affirmatively compensated in this society. Men's physiology defines most sports, their needs define auto and health insurance coverage, their socially-designed biographies define workplace expectations and successful career patterns, their perspectives and concerns define quality in scholarship, their experiences and obsessions define merit, their objectification of life defines art, their military service defines citizenship, their presence defines family, their inability to get along with each other—their wars and rulerships—defines history, their image defines god, and their genitals define sex. For each of their differences from women, what amounts to an affirmative action plan is in effect, otherwise known as the structure and values of American society. (Mackinnon 1987: 36)

All of this is 'gender-neutral', in the sense that women are not arbitrarily excluded from pursuing the things society defines as valuable. But it is sexist, because the things being pursued in a gender-neutral way are based on men's interests and values. Women are disadvantaged, not because chauvinists arbitrarily favour men in the awarding of jobs, but because the entire society systematically favours men in the defining of jobs, merit, etc.

Indeed, the more society defines positions in a gendered way, the less the difference approach is able to detect an inequality. Consider a society which restricts access to contraception and abortion, which defines paying jobs in such a way as to make them incompatible with child-bearing and child-rearing, and which does not provide economic compensation for domestic labour. Every woman who faces an unplanned pregnancy, and who cannot both raise children and work for wages, is rendered economically dependent on someone who is a stable income-earner (i.e. a man). In order to ensure that she acquires this support, she must become sexually attractive to men. Knowing that this is their likely fate, many girls do not try as hard as boys to acquire employment skills which can only be exercised by those who avoid pregnancy. Where boys pursue personal security by increasing their employment skills, girls pursue security by increasing their attractiveness to men. This, in turn, results in a system of cultural identifications in which masculinity is associated with income-earning, and femininity is defined in terms of sexual and domestic service for men, and the nurturing of children. So men and women enter marriage with different income-earning potential, and this disparity widens during marriage, as the man acquires valuable job experience. Since the woman faces greater difficulty supporting herself outside of the marriage, she is more dependent on maintaining the marriage, which allows the man to exercise greater control within it.

In such a society, men as a group exercise control over women's general life-chances (through political decisions about abortion, and economic decisions concerning job requirements), and individual men exercise control over economically vulnerable women within marriages. Yet there need be no arbitrary discrimination. All of this is gender-neutral, in that one's gender does not necessarily affect how one is treated by those in charge of distributing contraception, jobs, or domestic pay. But whereas the difference approach takes the absence of arbitrary discrimination as evidence of the absence of sexual inequality, it may in fact be evidence of its pervasiveness. It is precisely because women are dominated in this society that there is no need for them to be discriminated against. Arbitrary discrimination in employment is not only unnecessary for the maintenance of male privilege, it is unlikely to occur, for most women will never be in a position to be arbitrarily discriminated against in employment. Perhaps the occasional woman can overcome the social pressures supporting traditional sex-roles. But the greater the domination, the less the likelihood that any women will be in a position to compete for employment, and hence the less room for arbitrary discrimination. The more sexual inequality there is in society, the more that social institutions reflect male interests, the less arbitrary discrimination there will be.

None of the contemporary Western democracies correspond exactly to this model of a patriarchal society, but they all share some of its essential features. And if we are to confront these forms of injustice, we need to reconceptualize sexual inequality as a problem, not of arbitrary discrimination, but of domination. As Mackinnon puts it,

to require that one be the same as those who set the standard—those which one is already socially defined as different from—simply means that sex equality is conceptually designed never to be achieved. Those who most need equal treatment will be the least similar, socially, to those whose situation sets the standard as against which one's entitlement to be equally treated is measured. Doctrinally speaking, the deepest problems of sex inequality will not find women 'similarly situated' to men. Far less will practices of sex inequality require that acts be intentionally discriminatory. (Mackinnon 1987: 44; cf. Taub and Schneider 1982: 134)

The subordination of women is not fundamentally a matter of irrational differentiation on the basis of sex, but of male supremacy, under which gender differences are made relevant to the distribution of benefits, to the systematic disadvantage of women (Mackinnon 1987: 42; Frye 1983: 38).

Since the problem is domination, the solution is not only the absence of discrimination, but the presence of power. Equality requires not only equal opportunity to pursue male-defined roles, but also equal power to create female-defined roles, or to create androgynous roles that men and women have an equal interest in filling. The result of such empowerment could be very different from our society, or from the equal-opportunity-to-enter-male-institutions society that is favoured by contemporary sex-discrimination theory. From a position of equal power, we would not have created a system of social roles that defines 'male' jobs as superior to 'female' jobs. For example, the roles of male and female health practitioners were redefined by men against the will of women in the field. With the professionalization of medicine, women were squeezed out of their traditional health care roles as midwives and healers, and relegated to the role of nurse—a position which is subservient to, and financially less rewarding than, the role of doctor. That redefinition would not have happened had women been in a position of equality, and will have to be rethought now if women are to achieve equality.

Acceptance of the dominance approach would require many changes in gender relations. But what changes would it require in our theories of justice? Most of the theorists discussed in previous chapters implicitly or explicitly accept the difference approach. But does that reflect a flaw in their principles, or a flaw in the way those principles have been applied to issues of gender? Many feminists argue that the flaw lies in the principles themselves, that 'male-stream' theorists (as Mary O'Brien calls them) on both the right and left interpret equality in ways that are incapable of recognizing women's subordination. Indeed, some feminists argue that the struggle against sexual subordination requires us to abandon the very idea of interpreting justice in terms of equality. Elizabeth Gross argues that since women must be free to redefine social roles, their aims are best described as a politics of 'autonomy' rather than a politics of 'equality':

Autonomy implies the right to see oneself in whatever terms one chooses— which may imply an integration or alliance with other groups and individuals or may not. Equality, on the other hand, implies a measurement according to a given standard. Equality is the equivalence of two (or more) terms, one of which takes the role of norm or model in unquestionable ways. Autonomy, by contrast, implies the right to accept or reject such norms or standards according to their appropriateness to one's self-definition. Struggles for equality . . . imply an acceptance of given standards and a conformity to their expectations and requirements. Struggles for autonomy, on the other hand, imply the right to reject such standards and create new ones. (Gross 1986: 193)

Gross assumes that sex equality must be interpreted in terms of eliminating arbitrary discrimination. But the dominance approach is also an interpretation of equality, and if we accept it, then autonomy becomes a part of the best theory of sexual equality, not a competing value. The argument for women's autonomy appeals to, rather than conflicts with, the deeper idea of moral equality, for it asserts that women's interests and experiences should be equally important in shaping social life. As Zillah Eisenstein puts it, 'equality in this sense means individuals' having equal value as human beings. In this vision equality does not mean to be like men, as they are today, or to have equality with one's oppressors' (Eisenstein 1984: 253).

So the dominance approach shares with mainstream theorists the commitment to equality. But is it consistent with the way mainstream theorists interpret that commitment—for example, does the dominance approach fit into the liberal view of equality of resources? Mackinnon argues that the dominance approach takes us beyond the basic principles of liberalism. Is this so? It is true that liberal theorists, like other malestream theorists, have accepted the difference approach to sex equality, and, as a result, have not seriously attacked women's subordination. But one can argue that liberals (and other contemporary theorists) are betraying their own principles in adopting the difference approach.[3] Indeed the disjunction between the difference approach and liberal principles is obvious. Liberalism's commitment to autonomy and equal opportunity, and to an ambition-sensitive, endowment-insensitive distribution of resources, rules out traditional gender divisions. There seems to be no reason why the gender-bias of existing social roles would not be recognized by the contractors in Rawls's original position as a source of injustice. While Rawls himself says nothing about how his contractors would interpret sex equality, others have argued that the logic of Rawls's construction—i.e. the commitment to eliminating undeserved inequalities, and to the freedom to choose our ends—requires radical reform. For example, Karen Green argues that the contractors' interest in equal liberty requires redistributing domestic labour (Green 1986: 31–5). And Susan Okin argues that Rawls's contractors would insist on a more complete attack on the system of gender differentiation, eliminating both the unequal domestic division of labour and sexual objectification (Okin 1987: 67–8). Similar conclusions about the injustice of traditional gender-roles can be reached if we ask whether these roles pass Dworkin's test of fairness (cf. ch. 3, s. 5 above).

While liberals have historically endorsed the difference approach, this is the product, not of flawed principles, but of flawed applications of those principles. This is not to say that liberals have simply overlooked the problem of sexual inequality, as if by accident. There are obvious reasons of self-interest why male theorists have avoided the dominance approach. Moreover, as we will see, a commitment to this stronger account of sexual equality raises difficult questions about the relationship between public and private, and between justice and care.

2. THE PUBLIC AND THE PRIVATE

If we employ the dominance approach to sex equality, one of the central issues concerns the unequal distribution of domestic labour, and the relationship between family and workplace responsibilities. But mainstream theorists have been wary of confronting family relations and judging them in the light of standards of justice. Classical liberals, for example, assumed that the (male-headed) family is a biologically determined unit, and that justice only refers to the conventionally determined relations between families (Pateman 1980: 22–4). Hence the natural equality they discuss is of fathers as representatives of families, and the social contract they discuss governs relations between families. Justice refers to the 'public' realm, where adult men deal with other adult men in accordance with mutually agreed-upon conventions. Familial relationships, on the other hand, are 'private', governed by natural instinct or sympathy.

Contemporary theorists deny that only men are capable of acting within the public realm. But while sexual equality is now affirmed, this equality is still assumed, as in classical liberal theory, to apply to relations outside the family. Theorists of justice continue to ignore relations within the family, which is assumed to be an essentially natural realm. And it is still assumed, implicitly or explicitly, that the natural family unit is the traditional male-headed family, with women performing the unpaid domestic and reproductive work. For example, while J. S. Mill emphasized that women were equally capable of achievement in all spheres of endeavour, he assumed that women would continue to do the domestic work. He says that the sexual division of labour within the family is 'already made by consent, or at all events not by law, but by general custom', and he defends this as 'the most suitable division of labour between the two persons':

Like a man when he chooses a profession, so, when a woman marries, it may in general be understood that she makes choice of the management of a household, and the bringing up of a family, as the first call upon her exertions, during as many years of her life as may be required for the purpose; and that she renounces, not all other objects and occupations, but all which are not consistent with the requirements of this. (Mill and Mill 1970: 179)

While contemporary theorists are rarely as explicit as Mill, they implicitly share his assumption about women's role in the family (or if they do not, they say nothing about how domestic labour should be rewarded or distributed). For example, while Rawls says that the family is one of the social institutions to be evaluated by a theory of justice, he simply assumes that the traditional family is just, and goes on to measure just distributions in terms of the 'household income' which accrues to 'heads of households', so that questions of justice within the family are ruled out of court.[4] The neglect of the family has even been present in much of liberal feminism, which 'accepted the division between the public and private spheres, and chose to seek equality primarily in the public realm' (Evans 1979: 19).

The limits of any approach to sex equality that neglects the family have become increasingly clear. As we have seen, the result of women's 'double-day' of work is that women are concentrated in low-paying, part-time work, which in turn makes them economically dependent. But even if this economic vulnerability were removed, by guaranteeing an annual income to everyone, there is still the injustice that women are presented with a choice between family and career that men do not face. Mill's claim that a woman who enters a marriage accepts a full-time occupation, just like a man entering a profession, is strikingly unfair. After all, men also enter the marriage—why should marriage have such different and unequal consequences for men and women? The desire to be a part of a family should not preclude one's having a career, and in so far as it does have unavoidable consequences for careers, they should be borne equally by men and women.

Moreover, there remains the question of why domestic labour is not given greater public recognition. Even if men and women share the unpaid domestic labour, this would hardly count as genuine sexual equality if the reason why it was unpaid was that our culture devalues 'women's work', or anything 'feminine'. Sexism can be present not only in the distribution of domestic labour, but also in its evaluation. And since the devaluation of housework is tied to the broader devaluation of women's work, then part of the struggle for increased respect for women will involve increased respect for their contribution to the family. The family is therefore at the centre of both the cultural devaluation and economic dependence which attach to women's traditional roles. And the predictable result is that men have unequal power in nearly all marriages, power which is exercised in decisions concerning work, leisure, sex, consumption, etc., and which is also exercised, in a significant minority of marriages, in acts or threats of domestic violence (Okin 1989b: 128–30).

The family is therefore an important locus of the struggle for sexual equality. There is an increasing consensus amongst feminists that the fight for sex equality must go beyond public discrimination to the patterns of domestic labour and women's devaluation in the private sphere. In fact, Carole Pateman says that the 'dichotomy between the public and the private ... is, ultimately, what the feminist movement is all about' (Pateman 1987: 103).

Confronting the injustice of the private sphere would require substantial changes in family life. But what changes does it require of theories of justice? As we have seen, the failure to confront gender inequalities in the family can be seen as a betrayal of liberal principles of autonomy and equal opportunity. According to some feminist critics, however, liberals refuse to interfere in the family, even to advance liberal goals of autonomy and equal opportunity, because they are committed to a public–private distinction, and because they see the family as the centre of the private sphere. Thus Jaggar argues that because the liberal right to privacy 'encompasses and protects the personal intimacies of the home, the family, marriage, motherhood, procreation, and child rearing', any liberal proposals to interfere in the family in the name of justice 'represent a clear departure from this traditional liberal conception of the family as the center of private life. . . . As the liberal feminist emphasis on justice comes increasingly to overshadow its respect for so-called private life, one may begin to wonder whether the basic values of liberalism are ultimately consistent with one another' (Jaggar 1983: 199). In other words, liberals must give up either their commitment to sexual equality, or their commitment to the public–private distinction.

However, it is not clear that 'the traditional liberal conception' views the family 'as the center of private life'. There are in fact two different conceptions of the public–private distinction in liberalism: the first, which originated in Locke, is the distinction between the political and the social; the second, which arose with Romantic-influenced liberals, is the distinction between the social and the personal. Neither treats the family as wholly private, or explains or justifies its immunity from legal reform. Indeed, each distinction, if applied to the family, provides grounds for criticizing the traditional family. However, liberals have not applied these distinctions to the family, and have generally neglected the role of the family in structuring both public and private life.

(a) State and civil society

The first version of liberalism's public–private distinction is illustrated by Constant's distinction between ancient and modern freedom. The liberty of the ancients was their active participation in the exercise of political power, not the peaceful enjoyment of personal independence. The Athenians were free men because they were collectively self-governing, although they lacked personal independence and civil liberties, and were expected to sacrifice their pleasures for the sake of the *polis*. The liberty of the moderns, on the other hand, lies in the unimpeded pursuit of happiness in their personal occupations and attachments, which requires freedom from the exercise of political power. Whereas the ancients sacrificed private liberty to promote political life, moderns view politics as a means (and somewhat of a sacrifice) needed to protect their private life. Liberalism expresses its commitment to modern liberty by sharply separating the public power of the state from the private relationships of civil society, and by setting strict limits on the state's ability to intervene in private life.

Critics have often objected to this public–private distinction on the grounds that liberalism's emphasis on private life is antisocial. According to Marx, for example, the individual rights emphasized by liberals are the freedoms of 'a man treated as an isolated monad and withdrawn into himself . . . the right of man to freedom is not based on the union of man with man, but on the separation of man from man. It is the right to this separation' (Marx 1977*b*: 53). But the liberal view actually presupposes our natural sociability. As Nancy Rosenblum notes,

> this boundary between spheres does not imply that private life is radically apolitical or antisocial. Private life means life in civil society, not some presocial state of nature or antisocial condition of isolation and detach- ment . . . private liberty provides escape from the surveillance and interference of public officials, multiplying possibilities for private associ- ations and combinations. . . . Far from inviting apathy, private liberty is supposed to encourage public discussion and the formation of groups that give individuals access to wider social contexts and to government. (Rosenblum 1987: 61)

When the state leaves people in the 'perfect independence' of private life, it does not leave them in isolation, but rather leaves them free to form and maintain 'associations and combinations', or what Rawls calls 'free social unions'. Because we are social animals, individuals will use their freedom to join with others in the pursuit of shared ends. Freedom, for classical liberals, was indeed based on the 'union of man with man', but they believed that the union of men arising from free association in civil society is more genuine, and more free, than the coerced unity of political associations. The liberal ideal of private life was not to protect the individual from society, but to free society from political interference. It is more accurate to view liberalism, not as antisocial, but as 'the glorification of society', for liberals 'rated social life the highest form of human achievement and the vital condition for the development of morality and rationality', while the political was reduced to 'the harsh symbol of the coercion necessary to sustain orderly social transactions' (Wolin 1960: 363, 369, 291; cf. Holmes 1989: 248; Schwartz 1979: 245).

The basic issue in evaluating this version of the public–private split is not how much individuals require society for their freedom, but how much social individuals need the state for their freedom. As we saw in chapter 6, this has been obscured by communitarian critiques of liberal 'atomism' (ch. 6, s. 4 above). But when Aristotle said that men were *zoon politikon*, he did not mean simply that men are social animals. On the contrary, 'the natural, merely social companionship of the human species was considered to be a limitation imposed upon us by the needs of biological life, which are the same for the human animal as for other forms of animal life' (Arendt 1959: 24). Political life, on the other hand, was different from, and higher than, our merely social life.

Various attempts have been made to overturn the liberal glorification of society, and to reinstate politics as a higher form of life. But the liberal view pervades the modern age, and it is implicitly shared by even its most radical critics (Wolin 1960: 290, 414–16). Whereas the Greeks felt that 'under no circumstances could politics be only a means to protect society', modern theorists simply disagree on what kind of society politics should serve—is it 'a society of the faithful, as in the Middle Ages, or a society of property-owners, as in Locke, or a society relentlessly engaged in a process of acquisition, as in Hobbes, or a society of producers, as in Marx, or a society of job-holders, as in our own society, or a society of laborers, as in socialist and communist countries. In all these cases, it is the freedom . . . of society which requires and justifies the restraint of political authority. Freedom is located in the realm of the social, and force or violence becomes the monopoly of government' (Arendt 1959: 31). This is one of those cases, like the commitment to moral equality, where liberalism has simply won the historical debate, and all subsequent debate occurs, in a sense, within the boundaries of basic liberal commitments.

This is the first form of liberalism's public–private distinction. Feminists have objected to it on a number of grounds. The most pressing objection is that most liberal descriptions of the social realm make it sound as though it contains only adult (and able-bodied) men, neglecting the labour needed to create and nourish these participants, labour performed mainly by women, mainly in the family. As Pateman notes, 'liberalism conceptualizes civil society in abstraction from ascriptive domestic life', and so 'the latter remains "forgotten" in theoretical discussion. The separation between private and public is thus [presented] as a division *within* . . . the world of men. The separation is then expressed in a number of different ways, not only private and public but also, for example, "society" and "state"; or "economy" and "politics"; or "freedom" and "coercion"; or "social" and "political"' (Pateman 1987: 107), all of which are divisions 'within the world of men'.

Domestic life, therefore, has tended to fall outside both state and civil society. Why is the family excluded from civil society? The answer cannot be that it is excluded because it falls into the private realm, for the problem here is precisely that it is not viewed as part of the private realm, which is the realm of liberal freedom. This exclusion of the family is surprising, in one sense, for the family seems a paradigmatically social institution, potentially based on just the sort of co-operation that liberals admired in the rest of society, yet currently mired in just the sort of ascriptive restrictions which liberals abhorred in feudalism. Yet liberals who were concerned with protecting social life, and men's access to it, have not been concerned with ensuring either that domestic life is organized along principles of equality and consent, or that domestic arrangements do not impede women's access to other forms of social life. Why did liberals, who opposed ascriptive hierarchy in the realm of science, religion, culture, and economics, show no interest in doing the same for the domestic sphere?[5]

The obvious explanation is that male philosophers had no interest in questioning a sexual division of labour from which they benefited. This was rationalized at the level of theory through the assumption that domestic roles are biologically fixed, an assumption grounded either in claims of women's inferiority, or in the more recent ideology of the sentimental family, which says that the sentimental tie which naturally arises between mother and child is incompatible with the character traits needed for social or political life (Okin 1981). But while most liberal theorists have invoked one or other of these assumptions, they are not distinctively liberal views, and there is no logical or historical connection between them and acceptance of the liberal state–society distinction.

The sad fact of the matter is that almost all political theorists in the Western tradition, whatever their views on the state–society distinction, have accepted one or other of these justifications for separating domestic life from the rest of society, and for relegating women to it. As Kennedy and Mendus note, 'In almost all respects the theories of Adam Smith and Hegel, of Kant and Mill, of Rousseau and Nietzsche are poles apart, but in their treatment of women, these otherwise diverse philosophers present a surprisingly united front'. Male theorists on all points of the political spectrum have accepted that 'the confinement of women to the private [domestic] sphere is justified by reference to women's particular-istic, emotional, non-universal nature. Since she knows only the bonds of love and friendship, she will be a dangerous person in political life, prepared, perhaps, to sacrifice the wider public interest to some personal tie or private preference' (Kennedy and Mendus 1987: 3–4, 10). In other words, liberals have endorsed the domestic–public distinction for the same reasons that anti-liberals endorsed it, not because they believe in a public–private distinc-tion.[6]

In fact, those theorists who have rejected the liberal public–private distinction have tended to sharpen the traditional domestic–public split. For example, while the ancient Greeks had no conception of the sort of private sphere which liberals favour, they did have a sharp distinction between the domestic household and the public realm which condemned women to public invisibility (Elshtain 1981: 22; Arendt 1959: 24; Kennedy and Mendus 1987: 6). Far from denying the domestic–public split, 'at the roof of Greek political consciousness we find an unequaled clarity and articulateness in drawing this distinction' (Arendt 1959: 37). Similarly, while Rousseau opposed the liberal separation of public and private, he presented his vision of an integrated society 'as though it were and should be entirely male, supported by the private female familial structure' (Eisenstein 1981: 77; cf. Elshtain 1981: 165; Pateman 1975: 464). Indeed, he endorsed the Greek view that when women married, 'they disappeared from public life; within the four walls of their home they devoted themselves to the

care of their household and family. This is the mode of life
prescribed for women alike by nature and reason' (Rousseau, in
Eisenstein 1981: 66). Finally, while Hegel rejected liberalism's
'radical separation' of the public and private spheres, his theory
'provides the most graphic example of the way the sentimental
domestic family has been used to define women's capacities, and to
justify their subordination, lack of education, and exclusion from
the public realms of the market, citizenship, and intellectual life'
(Elshtain 1981: 176; Okin 1981: 85).

So the liberal public–private distinction is different from the
domestic–public distinction. Are there any feminist grounds for
rejecting the liberal state–society distinction, once we distinguish
it from the domestic–public split? Many contemporary feminists
accept the essential features of the liberal view of the relationship
between state and society, and so between public and private in that
sense.[7] For one thing, the Greek elevation of the political is based
on a nature–culture dualism of just the sort that feminists have
argued is at the root of the cultural devaluation of women in our
society. One important strand in the devaluation of women's work,
particularly in bearing and rearing children, is the idea that it is
merely natural, a matter of biological instinct rather than cultural
knowledge. Thus women are associated with the merely animal
functions of domestic labour, whereas men achieve truly human
lives by choosing activities according to cultural goals, not natural
instincts.

The claim that politics is a higher form of life often rests on the
view that social life, like domestic life, is mired in 'the heterono-
mous realm of particular need, interest, and desire' (Young 1989:
253). According to Greek thought, social life remains 'in nature's
prescribed cycle, toiling and resting, laboring and consuming,
with the same happy and purposeless regularity with which day and
night and life and death follow each other' (Arendt 1959: 106).
This 'purposeless regularity' of everyday life is ultimately insignificant, destined to pass into the dust from which it came. Only
politics is citizens' 'guarantee against the futility of individual life'
(Arendt 1959: 56). Because politics attempts to transcend 'nature's
cycles', 'it was a matter of course that the mastering of the
necessities of life in the household was the condition for freedom
of the polis . . . household life exists for the sake of the "good life"
in the polis' (Arendt 1959: 30–31, 37). Indeed, 'no activity that

served only the purpose of making a living, of sustaining only the life process, was permitted to enter the political realm' (Arendt 1959: 37). It is difficult to imagine a conception of public life in sharper opposition to Adrienne Rich's account of women's work as 'world-protection, world-preservation, world-repair . . . the invisible weaving of a frayed and threadbare family life' (Rich 1979: 205).

Moreover, since the priority of politics over society often rests on its alleged universality or commonality, protection of this universality requires separating politics from the realm of particularity, and that has invariably meant separating it from domestic concerns. As Iris Young notes,

in extolling the virtues of citizenship as participation in a universal public realm, modern men expressed a flight from sexual difference. . . . Extolling a public realm of manly virtue and citizenship as independence, generality, and dispassionate reason entailed creating the private sphere of the family as the place to which emotion, sentiment, and bodily needs must be confined. The generality of the public thus depends on excluding women. (Young 1989: 253–4)

Unlike the Greeks who valued politics as the transcendence of nature, and the Hegelians who valued politics as the transcendence of particularity, feminists and liberals share a basic commitment to viewing public power as a means for the protection of particular interests and needs.

This does not show that feminists and liberals agree on all aspects of the relationship between state and society. Even if we agree that public power should be justified in terms of the promotion of private interests in civil society, there are many areas of potential disagreement. Firstly, we might think that social life is not as stable and self-adjusting as liberals suppose. For example, we might think that individuals will not, by themselves, maintain the web of social relationships passed down to them. They will opt in and out of all social ties with such dizzying rapidity that society will disintegrate unless the state actively intervenes to encourage social groups. This is the ultimate message of communitarians, a message that, despite their emphasis on human sociability, actually presupposes that people need to be guided by government into social life (ch. 6, s. 4*b* above). But it is a legitimate concern, and we may want government to encourage the maintenance of certain social ties, including familial ones, and make exit from those ties more difficult.

Secondly, we might question liberalism's faith that if everyone has free and fair access to the means of expression and association, then truth will win out over falsity, and understanding over prejudice, without governments having to monitor these cultural developments. Liberals tend to believe that cultural oppression cannot survive under conditions of civil freedom and material equality. But there may be some false and pernicious cultural representations that are invulnerable to social criticism, that survive and even flourish in a free and fair fight with truth. Pornography, and other cultural representations of women, are an example. Liberals believe that if pornography does not harm women, then the falseness of its representation of sexuality is not grounds for restricting it, not because ideas are powerless, but because freedom of speech and association in civil society is a better testing ground for ideas than the coercive apparatus of the state. To some people, this will seem an unwarranted naïvety about the power of free speech in civil society to weed out cultural oppression. As Mackinnon puts it, if free speech helps discover truth, 'why are we now—with more pornography available than ever before—buried in all these lies?' (Mackinnon 1987: 155). She argues that this faith in free speech shows that 'liberal morality cannot deal with illusions that *constitute* reality' (Mackinnon 1987: 162).

While these areas of possible dispute between liberals and feminists are of the first importance (and involve some of the empirical questions about state and culture that I raised at the end of the last chapter), they are located within a shared commitment to the priority of social life over politics.

(b) The personal and the social: the right to privacy

The original liberal public–private split has been supplemented in the last hundred years by a second distinction, one which separates the personal or the intimate from the public, where the 'public' includes both state and civil society. This second distinction arose primarily amongst Romantics, not liberals, and indeed arose partly in opposition to the liberal glorification of society. Whereas classical liberals emphasized society as the basic realm of personal freedom, Romantics emphasized the effects of social conformity on individuality. Individuality was threatened not only by political coercion, but also by the seemingly omnipresent pressure of social expectations. For romantics, 'private' means detachment from mundane existence, [and] is associated with self-development, self-expression, and artistic creation. . . . In classical liberal thought, by contrast, 'private' refers to society, not personal retreat, and society is a domain of free rational activity rather than expressive licence. Liberalism protects this sphere by restricting the exercise of government power and by enumerating civil liberties. Pure romanticism and conventional liberalism are separated not only by their notions of private life, but also by their motivations for designating a privileged private sphere. (Rosenblum 1987: 59)

Romantics included social life in the public realm because the bonds of civil society, while non-political, still subject individuals to the judgement and possible censure of others. The presence of others can be distracting, disconcerting, or simply tiring. Individuals need time for themselves, away from public life, to contemplate, experiment with unpopular ideas, regenerate strength, and nurture intimate relationships. In these matters, social life can be just as demanding as political life. In fact, 'modern privacy in its most relevant functions, to shelter the intimate, was discovered as the opposite not of the political sphere but of the social' (Arendt 1959: 38; cf. Benn and Gaus 1983: 53). Hence Romantics viewed 'every formal association with others except for intimate relations like friendship or love' as public (Rosenblum 1987: 67).

While this second public–private distinction arose in opposition to liberalism, modern liberals have accepted much of the Romantic view, and have tried to integrate its emphasis on social pressures with the classical liberal emphasis on social freedom. The Romantic emphasis on privacy in fact coincided with liberal fears about the coercive power groups exercised over their own members in professional associations, labour unions, educational institutions, etc., and about the more generalized pressure for social uniformity, against which the plurality of associations and the marketplace of ideas provided inadequate protection for individuality. As a result, modern liberalism is concerned not only to protect the private sphere of social life, but also to carve out a realm *within the private sphere* where individuals can have *privacy*. Private life, for liberals, now means both active involvement in the institutions of civil society, as classical liberals emphasized, and personal retreat from that ordered social life, as Romantics emphasized.[8]

This second form of liberalism's public–private distinction is often discussed under its legal guise of a 'right to privacy'. Like the first public–private distinction, it has become the target of feminist criticism. The decision which gave the right to privacy constitutional status in the United States, *Griswold* v *Connecticut* (381 US 479 [1965]), was initially seen as a victory for women, since it ruled that laws which denied access to contraception to married women violated the right of privacy. But it has since become clear that this right, as interpreted by the American Supreme Court, can also be a hindrance to further reform of women's domestic oppression. The idea of a right to privacy has been interpreted to mean that any outside interference in the family is a violation of privacy. As a result, it has served to immunize the family from reforms designed to protect women's interests—for example, state intervention which would protect women against abuse, or empower women to sue for non-support, or officially recognize the value of domestic labour (Taub and Schneider 1982: 122). The right to privacy 'reinforces the division between public and private that . . . keeps the private beyond public redress and depoliticizes women's subjection within it' (Mackinnon 1987: 102).

Hence this second public–private distinction has reinforced the tendency to exempt family relations from the test of public justice. But there is something unusual about the Supreme Court's interpretation of the right to privacy, for it defines individual privacy in terms of the collective privacy of the family. The right to privacy has been held to attach to families as units, not to their individual members. As a result, individuals have no claim to privacy within the family. If two people enter a marriage, the right to privacy guarantees that the state will not interfere with the couple's domestic decisions. But if the woman has no privacy within her marriage to begin with, and no power in the making of those decisions, then this right of family privacy will not provide her with any individual privacy, and indeed it precludes the state from taking action to protect her privacy.

There are some cases where the Court has explicitly appealed to the woman's *individual* privacy, even within the family. But they seem to be the exception to the rule (Eichbaum 1979). Why have family relations not been subjected to the test of individual privacy? The answer cannot be that the family is viewed as the heart of private life, because the problem here is precisely that the notion of privacy which is applied elsewhere is not applied to family relations. As June Eichbaum puts it, the idea of family-based privacy contradicts the whole point of a right to privacy: 'a right of privacy which protects the interest of a collective unit, the family, at the expense of individual autonomy, ignores the human necessity for privacy altogether and necessarily obfuscates privacy's deeper meaning' (Eichbaum 1979: 368). Protecting the family from state intervention does not necessarily guarantee women (or children) a sphere for personal retreat from the presence of others, or from the pressure to conform to others' expectations.

Why has the Supreme Court interpreted privacy as family-based? The answer seems to lie in the lingering influence of pre-liberal ideas about the naturalness of the traditional family. This is evident in the long tradition of judicial defences of the sanctity of the family, of which the 'right to privacy' is just the latest instalment. The first defence of family-based privacy was the paterfamilias doctrine, under which 'the family household was conceived as an extension of the personality of the *pater familias*', so that 'intervening in a man's family affairs was an invasion of his personal private sphere . . . in essence no different from requiring him to take baths more often' (Benn and Gaus 1983: 38). Under this doctrine, women became the husband's property on marriage, and so ceased to be persons under the law, their interests defined by, and submerged in, the family, which was taken to be their natural position. With the gradual recognition of the rights of other members of the household, there were challenges to the father's authority. But the legitimation of the traditional family provided by the paterfamilias doctrine was reaffirmed by conservative courts through a doctrine of 'family autonomy' in the 1920s. While the household was not the father's property, the basic structure of the traditional family remained immune to judicial reform because it was seen as a bastion of civilization, and a precondition for social stability (e.g. *Meyer* v *Nebraska*, 262 US 390 [1923]).

With the changing view of the family in the 1960s, the family autonomy doctrine in turn was challenged, and the court needed a new justification for leaving the family alone. The emerging emphasis on privacy was a tempting replacement, for the liberal concern with individual intimacy partially overlapped with the conservative concern with family autonomy, and provided some modern legitimacy for that old policy. But the change is more cosmetic than real, for what the Court means by privacy is remarkably similar to what was previously meant by paterfamilias or family autonomy.[9] Indeed, the American Supreme Court has not denied that its family-based right to privacy is a continuation of the old family autonomy doctrine. The Court has justified its emphasis on marital privacy by stressing 'the ancient and sacred character of marriage as the basis of their decisions' (Grey 1980: 84–5; cf. Eichbaum 1979: 372). Conversely, the Court has denied even the most basic components of a liberal conception of individual privacy if they are not tied to the traditional family structure—for example, the Supreme Court continues to uphold laws criminalizing homosexual relations between consenting adults in their own homes, and to deny that these laws are a violation of anyone's right to privacy (*Bowers* v *Hardwick*, 478 US 186 [1986]).

Thus the Romantic ideal of privacy came into the law fused with the conservative ideal of the heterosexual, officially organized family as a bastion of society. While the court invokes the language of a liberal public–private distinction, it is in fact invoking an illiberal public–domestic distinction, one which subordinates individual privacy to family autonomy. Mackinnon notes that

It is probably not coincidence that the very things feminism regards as central to the subjection of women—the very place, the body; the very relations, heterosexual; the very activities, intercourse and reproduction; and the very feelings, intimate—form the core of what is covered by privacy doctrine. From this perspective, the legal concept of privacy can and has shielded the place of battery, marital rape, and women's exploited labor; has preserved the central institutions whereby women are *deprived* of identity, autonomy, control and self-definition. . . . This right to privacy is a right of men 'to be let alone' to oppress women one at a time. . . . It keeps some men out of the bedrooms of other men. (Mackinnon 1987: 101–2)

The reason it is not coincidental that the right to privacy has immunized the domestic sphere is not that liberal privacy entails protecting domesticity, but rather that the protectors of domesticity have adopted the language of liberal privacy.

Once it is detached from patriarchal ideas of family autonomy, I believe that most feminists share the basic liberal motivations for respecting privacy—i.e. the value of having some freedom from distraction and from the incessant demands of others, and the value of having room to experiment with unpopular ideas and to nourish intimate relationships (Allen 1988). (Consider Virginia Woolf's well-known claim that every women should have 'a room of her own'.) In any event, liberalism's conception of privacy, like its state–society distinction, is not a defence of the domestic–public split. For intimacy needs defending outside of the family, and solitude needs to be defended within the family. The line between privacy and non-private, therefore, cuts across the domestic–public distinction. While we hope that the family forms a 'realm of privacy and personal retreat', for many people the family is itself an institution from which they desire privacy, and state action may be needed within the domestic sphere to protect privacy and prevent abuse.[10]

Given the centrality of the family to the system of sexual inequality, it is crucially important that theories of justice pay attention to the effects of family organization on women's lives. The refusal of mainstream theories to do this is often explained by saying that the family has been relegated to the private realm. But in a sense this underestimates the problem. The family has not so much been relegated to the private realm, as simply ignored entirely. And women's interests are harmed by the failure of political theory to examine the family in either its public or private components. For the gender-roles associated with the traditional family are in conflict not only with public ideals of equal rights and resources, but also with the liberal understanding of the conditions and values of private life.

3. AN ETHIC OF CARE

One consequence of the public–domestic distinction, and of the relegation of women to the domestic sphere, is that men and women have become associated with different modes of thought and feeling. Throughout the history of Western philosophy, we find political theorists distinguishing the intuitive emotional particularistic dispositions said to be required for women's domestic life from the rational impartial and dispassionate thought said to be required for men's public life. Morality

is fragmented into a 'division of moral labor' along the lines of gender. . . . The tasks of governing, regulating social order, and managing other 'public' institutions have been monopolized by men as their privileged domain, and the tasks of sustaining privatized personal relationships have been imposed on, or left to, women. The genders have thus been conceived in terms of special and distinctive moral projects. Justice and rights have structured male moral norms, values, and virtues, while care and responsiveness have defined female moral norms, values, and virtues. (Friedman 1987a: 94)

These two 'moral projects' have been viewed as fundamentally different, indeed conflicting, such that women's particularistic dispositions, while functional for family life, are seen as subversive of the impartial justice required for public life. Hence the health of the public has been said to depend on the exclusion of women (Okin 1990; Pateman 1980).

Because this contrast has historically been used to justify patriarchy, early feminists like Mary Wollstonecraft argued that women's particularistic emotional nature was simply the result of the fact that women were denied the opportunity to develop their rational capacities fully. If women thought only of the needs of the people around them, ignoring the needs of the general public, it was because they were forcibly prevented from accepting public responsibilities (Pateman 1980: 31). Some contemporary feminists argue that the whole tradition of distinguishing 'masculine' and 'feminine' morality is a cultural myth that has no empirical basis. But there is a significant strand of contemporary feminism which argues that we should take seriously women's different morality— we should view it as a mode of moral reasoning, not simply intuitive feeling, and as a source of moral insight, not simply the artificial result of sexual inequality. Where male theorists claimed that women's dispositions were intuitive in nature and private in scope, some feminists argue that they are rational and potentially public in scope. The particularistic thought women employ is a better morality than the impartial thought men employ in the public sphere, or at least a necessary complement to it, especially once we recognize that sex equality requires a breaking down of the public–domestic dichotomy.

The renewed feminist interest in women's modes of moral reasoning largely stems from Carol Gilligan's studies of women's moral development. According to Gilligan, men's and women's moral sensibilities do in fact tend to develop differently. Women tend to reason in a 'different voice', which she summarizes this way:

In this conception, the moral problem arises from conflicting responsibilities rather than from competing rights and requires for its resolution a mode of thinking that is contextual and narrative rather than formal and abstract. This conception of morality as concerned with the activity of care centers moral development around the understanding of responsibility and relationships, just as the conception of morality as fairness ties moral development to the understanding of rights and rules. (Gilligan 1982: 19)

These two 'voices' have been characterized in terms of an 'ethic of care' and an 'ethic of justice', which, Gilligan claims, are 'fundamentally incompatible' (Gilligan 1986: 238).

There is some controversy as to whether this different voice really exists, and, if it does, whether it is significantly correlated with gender. Some people argue that while there are two distinct moral voices of care and justice, men and women tend to employ both with roughly equal regularity. Others argue that while men and women often talk with a different voice, this obscures an underlying commonality: 'The moralization of gender is more a matter of how we *think* we reason than of how we actually reason'. We '*expect* women and men to exhibit this moral dichotomy', and, as a result, '*whatever* moral matters men concern themselves with are categorized, estimably, as matters of "justice and rights", whereas the moral concerns of women are assigned to the devalued categories of "care and personal relationships"' (Friedman 1987a: 96; cf. Baier 1987a: 48). Perhaps men and women speak in a different voice, not because their actual thoughts differ, but because men feel they should be concerned with justice and rights, and women feel they should be concerned with preserving social relations.[11]

Whatever the empirical findings about gender differences, there remains the question of whether there is a care-based approach to political questions that competes with justice, and, if there is, whether it is a superior approach. Some people have responded to Gilligan's findings by saying that the ethic of care, while a valid moral perspective, is not applicable outside the 'private' realm of friendship and family. It deals with the responsibilities we take on in virtue of participating in particular private relationships, rather than the obligations we owe to each other as members of the public (Kohlberg 1984: 358; Nunner-Winkler 1984). But many feminists argue that the care ethic, while initially developed in the context of private relationships, has public significance, and should be extended to public affairs.

What is the ethic of care? As is apparent in Gilligan's summary, there is more than one difference between the two moral voices. The differences can be looked at under three headings (cf. Tronto 1987: 648):

1. moral capacities: learning moral principles (justice) versus developing moral dispositions (care);
2. moral reasoning: solving problems by seeking principles that have universal applicability (justice) versus seeking responses that are appropriate to the particular case (care);
3. moral concepts: attending to rights and fairness (justice) versus attending to responsibilities and relationships (care).

I will look briefly at (1) and (2), before concentrating on (3), which I believe is the heart of the care–justice debate.

(a) Moral capacities

Joan Tronto says that the ethic of care 'involves a shift of the essential moral questions away from the question, What are the best principles? to the question, How will individuals best be equipped to act morally?' (Tronto 1987: 657). Being a moral person is less a matter of knowing correct principles, and more a matter of having the right dispositions—for example, the disposition to perceive people's needs accurately, and to come up with imaginative ways of meeting them.

It is true that most contemporary theorists of justice concentrate more on determining correct principles than on explaining how individuals become 'equipped to act morally'. But the former leads naturally to the latter, for the justice ethic also requires these moral dispositions. While justice involves applying correct principles, 'what it takes to bring such principles to bear on individual situations involves qualities of character and sensibilities which are themselves moral and which go beyond the straightforward process of consulting a principle and then conforming one's will and action to it' (Blum 1988: 485). Consider, for example, the dispositions required for jurors to decide whether someone used 'reasonable precautions' in negligence cases, or to decide when pay differentials between traditionally male and female jobs are 'discriminatory'. To act justly in these circumstances, sensitivity to historical factors and current possibilities is as important as 'the intellectual task of generating or discovering the principle' (Blum 1988: 486; cf.

Stocker 1987: 60). As we will see, there are some circumstances where it is important that principles of justice be easily interpreted, and their results easily predicted. But in many circumstances, moral sensitivities are required to see whether principles of justice are relevant to a situation, and to determine what those principles require. Hence justice theorists should join Gilligan in challenging the 'assumption that we need not worry what passions persons have, as long as their rational wills can control them' (Baier 1987*b*: 55). Even if justice involves applying abstract principles, people will only develop an effective 'sense of justice' if they learn a broad range of moral capacities, including the capacity for sympathetic and imaginative perception of the requirements of the particular situation.

Why have justice theorists neglected the development of the affective capacities underlying our sense of justice? Perhaps because the sense of justice grows out of a sense of care which is learned within the family. We could not teach children about fairness unless they had already learned within the family 'certain things about kindness and sensitivity to the aims and interests of others' (Flanagan and Jackson 1987: 635; cf. Baier 1987*a*: 42). Many justice theorists do recognize the role of the family in developing the sense of justice. Rawls, for example, has a lengthy discussion of how the sense of justice grows out of the moral environment of the family (Rawls 1971: 462–79). But this creates a contradiction within the justice tradition. As Okin puts it, 'in line with a long tradition of political philosophers', Rawls 'regards the family as a school of morality, a primary socializer of just citizens. At the same time, along with others in the tradition, he neglects the issue of the justice or injustice of the gendered family itself. The result is a central tension within the theory, which can be resolved only by opening up the question of justice within the family' (Okin 1989*a*: 230–1). Rawls begins his account of moral development by saying 'given that family institutions are just . . .' (Rawls 1971: 490). But, as we have seen, he does nothing to show that they are just. And 'if gendered family institutions are *not* just but are, rather, a relic of caste or feudal societies in which roles, responsibilities, and resources are distributed, not in accordance with the two principles of justice but in accordance with innate differences that are imbued with enormous social significance, then Rawls's whole structure of moral development seems to be built on uncertain ground' (Okin

1989*a*: 237; cf. Kearns 1983: 34–40). For example, what ensures that children are learning about equality rather than despotism, or reciprocity rather than exploitation? Investigating the justice of the family is important, therefore, not only as a site of sexual inequality, but also as a school for the sense of justice in boys and girls.

Rather than confront these questions, most theorists of justice have been content simply to assume that people have somehow developed the requisite capacities. But while they say little about this, they do recognize that 'to have failed to develop in oneself the capacity to be considerate of others is to have failed morally, if only because many duties simply cannot be carried out by a cold and unfeeling moral agent' (Sommers 1987: 78).

(b) Moral reasoning

So moral agents need 'the broader moral capacities' which Tronto discusses. But can these capacities take the place of principles? According to Tronto, the care ethic says that, *rather than* 'asserting moral principles', one's 'moral imagination, character, and actions must respond to the complexity of a given situation' (Tronto 1987: 657–8; cf. Baier 1987*a*: 40). In other words, these broader moral dispositions do not simply help individuals apply universal principles, they render such principles unnecessary, and perhaps counter-productive. We should construe morality in terms of attending to a particular situation, not in terms of applying universal principles. 'The idea of a just and loving gaze directed upon an individual reality . . . is the characteristic and proper mark of the moral agent', and this sort of 'ethical caring' does not depend 'on rule or principle' (Iris Murdoch, quoted in Grimshaw 1986: 234; cf. Ruddick 1980: 223–4; Noddings 1984: 81–94).

But what does it mean simply to attend to the situation? After all, not all contextual features are relevant to moral decisions. In making moral decisions, we do not simply attend to the different features of the situation, we also judge their relative significance. And while we want people to be good at attending to the complexity of the situation, we also want them to be good at identifying which features of the situation are the morally significant ones. And this seems to raise questions of principle rather than sensitivity: 'We have been told nothing about [the care ethic] until we are told what features of situations context-sensitive people pick out as morally salient, what weightings they put on these different features, and so on . . . we simply need to know more, in a detailed way, about to what and to whom women feel responsible, and about exactly what it is they care about' (Flanagan and Adler 1983: 592; Sher 1987: 180).

Ruddick claims that while we do distinguish salient and irrelevant features of moral situations, these distinctions come from the very process of attending to the situation, rather than from external principles. Someone who attends closely to a particular situation will come to see it as making demands on us. But while some moral considerations may be readily observable to anyone who has developed the capacity for sympathetic attention to a particular situation, there are other relevant considerations which are less obvious. For example, when are job qualifications discriminatory? As we have seen, the existing job situation may 'demand' someone who is free of child-care responsibilities, or who has a certain height or strength. Since these are genuinely relevant criteria for the job, it is only within a broader social perspective that we can see how their combined effect is to create a system of sexual inequality. In these circumstances, knowing when relevant criteria are none the less discriminatory, or when reverse discrimination is none the less legitimate, requires more than sympathetic attention to the particular situation. In order to know when there is a legitimate moral demand for affirmative action, we need to place the particular situation within a broader theory of sex equality.

Moreover, even if we have perceived all the relevant demands, these demands can conflict, and so detailed attention may lead to indecision in the absence of higher-level principles. If one is faced with a conflict between the demands of current male candidates and those of future generations of women, patient attention to the situation may just bring out how painful the conflict over affirmative action is. As Held notes, 'we have limited resources for caring. We cannot care for everyone or do everything a caring approach suggests. We need moral guidelines for ordering our priorities' (Held 1987: 119; Grimshaw 1986: 219).

Ruddick and Gilligan write as if appealing to principles involves abstracting from the particularity of the situation. But as Grimshaw notes, principles are not instructions to avoid examining the particulars, but rather are instructions about what to look for. Unlike 'rules', such as the ten commandments, which are intended as guides that can be applied without much reflection, a principle 'functions quite differently. It serves precisely to *invite* rather than block reflection', for it is 'a general consideration which one deems important to take into account when deciding what is the right thing to do' (Grimshaw 1986: 207–8). Every moral theory must have some account of such general considerations, and the sorts of considerations appealed to by theorists of justice often require, rather than conflict with, attention to particular details (Friedman 1987*b*: 203).

Some care theorists claim that the tendency to appeal to principles to adjudicate conflicts pre-empts the more valuable tendency to work out solutions in which the conflicts are overcome. For example, Gilligan claims that when constructing moral problems in terms of justice or care, her subjects either 'stood back from the situation and appealed to a rule or principle for adjudicating the conflicting claims or they entered the situation in an effort to discover or create a way of responding to all of the needs' (Gilligan 1987: 27). And indeed she cites many cases where girls were able to find a solution that responded to all of the needs in the particular situation, a solution which boys missed in their haste to find a principled adjudication of the conflict. But there will not always be a way to accommodate conflicting demands, and it is not clear that we should always try to accommodate all demands. Consider the demands of racist or sexist codes of honour. These are clear 'demands', but many of them are illegitimate. The fact that white men expect to be treated in a deferential way is no reason to accommodate such expectations. Even if we could accommodate them, we might provoke a conflict in order to make clear our disapproval. If we are to question these demands, then 'attention cannot always *just* be focused on the details and nuances of the particular situation', but rather must situate those details within some larger framework of normative principles (Grimshaw 1986: 238; Wilson 1988: 18–19).

(c) Moral concepts

The question, then, is not whether we need principles, but whether they should be principles that attend to 'rights and fairness' or 'responsibilities and relationships'. There are at least three different ways of construing the difference between these moral concepts:

1. universality versus concern for particular relationships;
2. respect for common humanity versus respect for distinct individuality;
3. claiming rights versus accepting responsibilities.

I will look at these in turn.

(i) Universality versus preserving relationships

One common way of distinguishing care and justice is to say that justice aims at universality or impartiality, whereas care aims at preserving the 'web of ongoing relationships' (Blum 1988: 473; Tronto 1987: 660). As Gilligan puts it, 'From a justice perspective, the self as moral agent stands as the figure against a ground of social relationships, judging the conflicting claims of self and others against a standard of equality or equal respect (the Categorical Imperative, the Golden Rule). From a care perspective, the relationship becomes the figure, defining self and others. Within the context of relationship, the self as a moral agent perceives and responds to the perception of need' (Gilligan 1987: 23). Hence, for Gilligan, 'morality is founded in a sense of concrete connection and direct response between persons, a direct sense of connection which exists prior to moral beliefs about what is right or wrong or which principles to accept. Moral action is meant to express and to sustain those connections to particular other people' (Blum 1988: 476).

There is some ambiguity in the notion of the 'existing web of relationships'. On one view, this refers to historically rooted relationships with particular others. If interpreted this way, however, the care ethic runs the danger of excluding the most needy, since they are most likely to be outside the web of relationships. Many care theorists recognize this danger. Tronto says that 'in focusing on the preservation of existing relationships, the perspective of care has a conservative quality', and that how to ensure 'that the web of relationships is spun widely enough so that some are not beyond its reach remains a central question. Whatever the weaknesses of Kantian universalism, its premiss of the equal moral worth and dignity of all humans is attractive because it avoids this problem' (Tronto 1987: 660–1). But the question is not simply to explain *how* 'social institutions might be arranged to expand these conventional understandings of the boundaries of care', but *why* they should be rearranged, unless we accept a universalistic principle of equal moral worth. Tronto's surprisingly tentative answer is that 'it may be possible to avoid the need for special pleading while at the same time stopping short of universal moral principles; if so, an ethic of care might be viable' (Tronto 1987: 661, 660).

Other care theorists, however, construe the 'existing web of relationships' in a more expansive way. Like Tronto, Gilligan says that 'each person is embedded within a web of ongoing relationships, and morality importantly if not exclusively consists in attention to, understanding of, and emotional responsiveness toward the individuals with whom one stands in these relationships' (Blum 1988: 473). But as Blum notes, 'Gilligan means this web to encompass all human beings and not only one's circle of acquaintances' (Blum 1988: 473). As one of the women in Gilligan's study puts it, we are responsible to 'that giant collection of everybody', so that 'the stranger is still another person belonging to that group, people you are connected to *by virtue of being another person*' (Gilligan 1982: 57, my emphasis; cf. 1982: 160). For Gilligan, what joins people in this giant web of relationships is not necessarily any direct interaction, but rather a shared humanity. Since Gilligan's conception of the web of relationships already includes everyone, her commitment to preserving the web of relationships entails, rather than conflicts with, her claim that the motivation of the care ethic is 'that everyone will be responded to and included, that no one will be left alone or hurt' (Gilligan 1982: 63).

Of course, once care theorists say that each person is connected to us 'by virtue of being another person', then it seems that they too are committed to a principle of universality. As soon as care and concern 'are detached from the demands of unique and historically rooted relationsips—as soon as they are said to be elicited merely by the affected parties' common humanity, or by the fact that those parties all have interests, or all can suffer', then 'we completely lose the contrast between the particularity of relationship and the generality of principle. Having lost it, we seem to be left with an approach that seeks to resolve moral dilemmas through sympathetic identification with all the affected parties.' And this sort of universality 'is at least closely related to that of the familiar impartial and benevolent observer' we find in Kantian and utilitarian theories (Sher 1987: 184). While Gilligan avoids the language of universality, her studies 'indicate that women's care and sense of responsibility for others are frequently universalized' (Okin 1990: 158; cf. Broughton 1983: 606; Kohlberg 1984: 356).

So the commitment to 'preserving the web of relationships' may or may not conflict with the commitment to universality, depending on how we interpret it. Much of the ethic of care literature has centred on the 'conflicted but creative tension' between the universalistic and more localized conceptions of our connection to others (Ruddick 1984: 239). Care theorists argue that we 'make moral progress . . . by expanding the scope of the injunctions to give care and to maintain connections' (Meyers 1987: 142), even though this requires 'transforming' and 'generalizing' some existing care practices (Ruddick 1980: 222, 226). But it is also true that 'the sense of responsibility at the core of the care perspective' tries to avoid 'imposing impartiality at the expense of ongoing attachment' (Meyers 1987: 142). It seems then that most care theorists accept Gilligan's commitment to a universalistic web of relations, but prefer to emphasize its continuity with Tronto's more localized web of relations. However, as Blum notes, 'how this extension to all persons is to be accomplished is not made clear' (Blum 1988: 473).[12]

(ii) Respect for humanity and respect for individuality

According to some care theorists, the problem with justice is not that it responds universally to all those who share our common humanity, but that it responds solely to people's common humanity, rather than to people's distinct individuality. Care theorists claim that for justice-based theorists, 'the moral significance of persons as the objects of moral concern is solely as bearers of morally significant but entirely general and repeatable characteristics' (Blum 1988: 475). Justice is concerned with the 'generalized other', and neglects the 'concrete other':

The standpoint of the generalized other requires us to view each and every individual as a rational being entitled to the same rights and duties we would want to ascribe to ourselves. In assuming the standpoint, we abstract from the individuality and concrete identity of the other. We assume that the other, like ourselves, is a being who has concrete needs, desires and affects, but that what constitutes his or her moral dignity is not what differentiates us from each other, but rather what we, as speaking and acting rational agents, have in common. . . . The standpoint of the concrete other, by contrast, requires us to view each and every rational being as an individual with a concrete history, identity and affective-emotional constitution. In assuming this standpoint, we abstract from what constitutes our commonality. . . . In treating you in accordance with the norms of friendship, love and care, I confirm not only your *humanity* but your human *individuality*. (Benhabib 1987: 87; cf. Meyers 1987: 146–7; Friedman 1987a: 105–10)

As Benhabib stresses, the standpoints of the general and concrete other are both fully universalized (indeed, she calls them 'substitutionalist universalism' and 'interactive universalism', respectively). But care, unlike justice, responds to our concrete differences, rather than our abstract humanity.

But this contrast seems overdrawn in both directions. Firstly the ethic of care, once universalized, also appeals to common humanity. As Sher notes, as soon as care and concern are 'said to be elicited merely by the affected parties' common humanity, or by the fact that those parties all have interests, or all can suffer', then they are 'viewed as appropriate responses to shared and repeatable characteristics' (Sher 1987: 184).

Secondly, theories of justice are not limited to respect for the generalized other. This is clear in the case of utilitarianism, which must attend to particularity in order to know whether a policy will promote people's various preferences. It may seem less clear in the case of Rawls's theory, and not surprisingly, many feminists point to his original position as a paradigm of justice-thinking. Because the original position requires individuals to abstract from their particular selves, it is said to exemplify a tradition in which 'the moral self is viewed as a *disembedded* and *disembodied* being' (Benhabib 1987: 81). But this misrepresents the original position. As Okin notes,

The original position requires that, as moral subjects, we consider the identities, aims, and attachments of every other person, however different they may be from ourselves, as of equal concern with our own. If we, who *do* know who we are, are to think *as if* we were in the original position, we must develop considerable capacities for empathy and powers of communicating with others about what different human lives are like. But these alone are not enough to maintain in us a sense of justice. Since we know who we are, and what are our particular interests and conceptions of the good, we need as well a great commitment to benevolence; to *caring* about each and every other as much as about ourselves. (Okin 1989a: 246)

Therefore, 'Rawls's theory of justice is itself centrally dependent upon the capacity of moral persons to be concerned about and to demonstrate care for others, especially others who are most different from themselves' (Okin 1989*a*: 247). Care theorists often say that conflict resolutions 'should be arrived at through the contextual and inductive thinking characteristic of taking the role of the particular other' (Harding 1987: 297). But this is precisely what the original position requires of us.

Benhabib questions whether 'taking the viewpoint of others' is truly compatible with reasoning behind a veil of ignorance, because justice is 'thereby identified with the perspective of the disembedded and disembodied generalized other. . . . The problem can be stated as follows: according to Kohlberg and Rawls, moral reciprocity involves the capacity to take the standpoint of the other, to put oneself imaginatively in the place of the other, but under conditions of the "veil of ignorance", the *other as different from the self*, disappears' (Benhabib 1987: 88–9; cf. Blum 1988: 475; Gilligan 1986: 240; 1987: 31). But this misrepresents how the original position operates. The fact that people are asked to reason in abstraction from their own social position, natural talents, and personal preferences when thinking about others does not mean that they must ignore the particular preferences, talents, and social position of others. And, as we have seen, Rawls insists that parties behind the original position must take these things into account (ch. 3, s. 3 above). Benhabib assumes that the original position works by requiring contractors to consider the interests of the other contractors (who all become 'generalized others' behind the veil of ignorance). But in fact the effect of the veil is that 'it no longer matters to the [contractor in] the original position who, if anyone, occupies the position with him or what its occupants' interests are. What matters to him are the desires and goals of every *actual* member of his society, because the veil forces him to reason *as if he were any one of them*' (Hampton 1980: 335). As we have seen, Hare's ideal sympathizer imposes the same requirement (ch. 2, s. 5*b* above). Both devices, impartial contractors and ideal sympathizers, work by requiring people to consider concrete others (cf. Broughton 1983: 610; Sher 1987: 184).[13]

(iii) Accepting responsibility and claiming rights

Since both ethics are universal, and both respect commonality as well as individuality, the difference (if there is one) lies elsewhere. One final distinction offered by Gilligan is that justice reasoning thinks of concern for others in terms of respecting rights-claims, whereas care reasoning thinks of concern for others in terms of accepting responsibilities. But what does this difference amount to? One central difference, according to Gilligan, is that accepting responsibility for others requires some positive concern for their welfare, whereas rights are essentially self-protection mechanisms that can be respected by simply leaving other people alone. Thus she equates talk about rights with individualism and selfishness, and says that rights-based duties to others are limited to reciprocal non-interference (Gilligan 1982: 22, 136, 147; cf. Meyers 1987: 146).

But this only holds for libertarian theories of rights. All of the other theories I have examined recognize positive duties concerning the welfare of others. So while the justice framework emphasizes people's rights, it is quite appropriate to say that these rights impose responsibilities on others. And indeed that is how some of Gilligan's respondents describe their ethic of care. For example, one woman says that 'People suffer, and that gives them certain rights, and that gives you a certain responsibility' (quoted in Broughton 1983: 605). It is true that some women 'think less about what they are entitled to than about what they are responsible for providing'. But they may regard themselves as responsible for providing care to others precisely because they regard others as being entitled to it— 'To suppose otherwise would be to conflate the well supported claim that women are less concerned than men with the protection of *their* rights with the quite different claim that women are less inclined than men to think that people *have* rights (or to hold views functionally equivalent to this)' (Sher 1987: 187).

Once we abandon the libertarian construal of rights as non-interference, the whole contrast between responsibilities and rights threatens to collapse (Okin 1990: 157). As Broughton puts it, 'Gilligan and her subjects seem to presuppose something like "the right of all to respect as a person", "the right to be treated sympathetically and as an equal", and "the duty to respect and not to hurt others"'. Hence 'it is difficult to see in what way she is not here recommending more or less binding rights and duties or perhaps even "principles" of personal welfare and benevolent concern' (Broughton 1983: 612). And while Gilligan insists that the two ethics are fundamentally different, she herself seems undecided about their relationship. She 'shifts between the ideas that the two ethics are incompatible alternatives to each other but are both adequate from a normative point of view; that they are complements of one another involved in some sort of tense interplay; and that each is deficient without the other and thus ought to be integrated' (Flanagan and Jackson 1987: 628). These shifts should not be surprising if, as I have argued, the key concepts Gilligan uses to distinguish the two ethics do not define genuine contrasts.[14]

While rights and responsibilities are not contrasting moral concepts, there is a difference in the kind of responsibility each ethic imposes on us. According to Sandra Harding, Gilligan's research shows that 'subjectively-felt hurt appears immoral to women whether or not it is fair', whereas men 'tend to evaluate as immoral only objective unfairness—regardless of whether an act creates subjective hurt' (Harding 1982: 237–8; 1987: 297). For example, men are less inclined to recognize any moral obligation to ameliorate subjective hurts that arise from someone's own negligence, since they are her own fault. Here there is a subjective hurt, but no objective unfairness, and so men tend to recognize no moral obligation. For women, on the other hand, the immorality of subjective hurts does not depend on the presence of objective unfairness.

There is a genuine contrast between taking subjective hurts or objective unfairness as the grounds for moral claims. Is this the fundamental difference between care and justice? It is certainly true that most justice theorists tie moral claims to objective unfairness rather than subjective hurt.[15] It is less clear whether the care ethic says that subjective hurts form the basis for moral claims, whether all subjective hurts, and only such hurts, ground moral claims. To care for somebody does not necessarily mean that one feels a moral obligation to attend to their every wish, or to spare them from all subjective hurts or disappointments. Care theorists have not in fact said much, so far, about how they understand the connection between subjective hurt, objective unfairness, and moral claims, and it is likely that different conceptions of ethical caring would arrive at different conclusions. So it is premature to assume that care and justice have fundamentally opposed views on this matter.

However, while the exact points of disagreement are unclear, it seems true that care theorists tend to emphasize subjective hurt rather than objective unfairness as the basis for moral claims. Before considering some of the reasons care theorists have for emphasizing subjective hurt, I will examine some of the reasons justice theorists have for preferring objective unfairness as the basis for moral claims. I will argue that the emphasis on objective unfairness, while initially plausible, is only legitimate in certain contexts—namely, interactions between competent adults. Indeed, it may be legitimate only when our interactions with competent adults are sharply separated from our interactions with dependants. If so, then the debate between care and justice reasoning becomes inextricably linked to the debate over the domestic–public distinction.

Why do justice theorists think it is important to limit our responsibility for others to the claims of fairness? If subjective hurts always give rise to moral claims, then I can legitimately expect, as a matter of ethical caring, that others attend to all of my interests. But for justice theorists, this ignores the fact that I should accept full responsibility for some of my own interests. In the justice perspective, I can legitimately expect, as a matter of fairness, that others attend to *some* of my interests, even if it limits the pursuit of their own good. But I cannot legitimately expect people to attend to all of my interests, for there are some interests which remain my own responsibility, and it would be wrong to expect others to forgo their good to attend to things which are my responsibility.

Consider someone who is generous with his time and money when his friends are in need, but is also exceedingly careless in his expenditures. As a result, he is often (unnecessarily) in need of help, and he relies on others to spare him from the consequences of his imprudence. Does he have a legitimate expectation that others help him—should we feel morally bound to spare him from the results of his carelessness? The subjective hurt approach says that we are irresponsible if we do not attend to his suffering. If he feels a subjective hurt, then we are required to attend to him, even though the hurt is the result of his own careless planning or extravagance. The justice ethic, however, says that he is irresponsible in expecting us to spare him from any suffering. His actions are his own responsibility, and it is immoral to make others pay for the costs of his carelessness.

Viewed this way, the debate between subjective hurts and objective unfairness is a genuine one, for there are importantly different positions we can take on the issue of responsibility for our own well-being. For care theorists, the emphasis on objective unfairness sanctions an abdication of moral responsibility, because it limits our responsibility for others to claims of unfairness, and thereby allows people to ignore avoidable suffering. For justice theorists, the emphasis on subjective hurts sanctions an abdication of moral responsibility, because it denies that the imprudent should pay for the costs of their choices, and thereby rewards those who are irresponsible, while penalizing those who act responsibly.

The debate between care and justice, therefore, is not between responsibility and rights. On the contrary, responsibility is central to the justice ethic. The reason why my claim on other people is limited to fairness is not that they have rights, but that I have responsibilities—part of my responsibility for others involves accepting responsibility for my own desires, and for the costs of my choices. As Rawls puts it, his theory 'relies on a capacity to assume responsibility for our ends' (Rawls 1982*b*: 169). Conversely, those who tie moral obligations to subjective hurts rather than objective unfairness must deny that we are responsible agents: they must argue 'that it is unreasonable, if not unjust, to hold [people] responsible for their preferences and to require them to make out as best they can' (Rawls 1982*b*: 168). Since Rawls thinks we have the capacity to assume this responsibility, his theory requires people to live within their means, to adjust their plans to the income they can rightfully expect. As a result, a careless and extravagant person cannot expect those who have been more responsible to pay for the costs of his imprudence: 'it is regarded as unfair that they now should have less in order to spare [him] from the consequences of [his] lack of foresight or self-discipline' (Rawls 1982*b*: 169). If we are obligated to spare people from all subjective hurts, then those who have responsibly attended to their own well-being will be asked to make continual sacrifices to aid those who have been irresponsibly careless or extravagant, and that is unfair.

The view that subjective hurts always give rise to moral claims is not only unfair, it can hide oppression. Subjective hurts are tied to expectations, and unjust societies create unjust expectations. Consider traditional marital relationships, in which 'men do not serve women as women serve men' (Frye 1983: 9, 10; cf. Friedman 1987*a*: 100–1; Grimshaw 1986: 216–19). Men expect women to attend to their needs, and so they feel subjective hurt whenever they are required to share the burdens of domestic life. Indeed, 'in all attempts to change exploitative or oppressive relations, someone is going to be deprived of something. They may be deprived of some attention, service or amenity to which they are accustomed. They may undergo some hardship or difficulty and experience this as lack of care' (Grimshaw 1986: 218). The oppressors will keenly feel any loss of privilege, while the oppressed are often socialized not to feel subjective hurt at their oppression. As a result, focusing on subjective hurts as the grounds for moral claims makes oppression harder to see. Within the justice perspective, on the other hand, the oppressors' subjective hurts have no moral weight, since they arise from unfair and selfish expectations. Claims of justice are determined by people's rightful expectations, not their actual expectations. (This explains why justice theorists say not only that subjective hurts lack moral significance in the absence of objective unfairness, but also that objective unfairness is immoral even when unaccompanied by subjective hurt, as when people are socialized to accept their oppression—cf. Harding 1987: 297.) In this sense, 'morally valid forms of caring and community presuppose prior conditions and judgements of justice' (Kohlberg 1984: 305).[16]

There is another problem with using subjective hurt as the basis of moral claims. While it imposes too little responsibility for our own well-being, it imposes too great a responsibility for others. If subjective hurt always calls forth a caring response, there seems to be nothing which limits our obligation to attend to others. There is always something more that we can do for others, if we attend closely enough to their desires—there is always some frustrated desire we can help fulfil. And this becomes self-reinforcing, for once someone knows that we are attending to them, they will come to expect attention, and then be even more hurt if our attention is withdrawn. As a result, the agent always faces moral claims on her time and energy, claims which leave no room for the free pursuit of her own projects.

So the idea that subjective hurts give rise to moral claims threatens both fairness and autonomy. Many care theorists recognize this problem, and try to put limits on what others can legitimately expect of us. Some theorists say that care-givers should attend to their own need for autonomy, or that genuine caring involves some kind of reciprocity, so that there are limits on how much others can expect of us without helping us in return (Ruddick

1984: 238; Gilligan 1982: 149; Noddings 1984: 105). In these and other ways, care theorists distance themselves from any simple equation of subjective hurt and moral claims.

But how much autonomy can we claim for ourselves, and how much reciprocity can we demand from others, without irresponsibly neglecting the subjective hurt of others? In line with their general methodology, care theorists say that the conflict between autonomy and responsibility for others must be decided contextually. Unlike one of Gilligan's male respondents, who said that we should treat this conflict as a 'mathematical equation' whose solution lies in a formula like 'one-fourth to the others and three-fourths to yourself' (Gilligan 1982: 35, 37), care theorists say we should judge the appropriateness of any demand for autonomy or reciprocity 'on the grounds of what is reasonable to expect from the individual being cared-for, along with what should be expected from such an individual given the nature of the caring relationship at hand' (Wilson 1988: 20). Care theorists, unlike justice theorists, do not try to resolve these issues by developing a comprehensive system of abstract rules that runs roughshod over the particularity of persons and their relationships.

However, this is one of the places where abstraction is an important virtue. If our aim is to ensure that the free pursuit of one's projects is not entirely submerged by the requirements of ethical caring, then we do not simply need limits on our moral responsibilities, we need *predictable* limits. We need to know *in advance* what we can rely on, and what we are responsible for, if we are to make long-term plans. It is not much good being told at the last minute that no one needs your moral help today, and that you are free to take a moral holiday, as it were. We can only take advantage of holidays if we can plan them, and that requires that we can determine *now* which interests we will be held responsible for *later*. And that, in turn, requires that when deciding at that later date who is responsible for attending to others, we do not make a fully context-sensitive decision.

For example, when my vacation day comes up at work, we do not ask who is least needed in the office. We ask whose turn it is under the system of rules. The result may be that some people will suffer the frustration of desires that a more contextual decision-making process would have fulfilled (other people in the office really would be less missed). But if we want to be able to make genuine commitments to our projects, then our claims must be insulated, to some extent, from the contingent desires of those around us. Abstract rules provide some security in the face of the shifting desires of others.

Of course, care theorists are right to say that some kinds of relationships must invoke different standards for balancing autonomy and responsibility. For example, we cannot expect children to have the same respect for autonomy and reciprocity as adults (I will return to this below). But for interactions between competent adults, an important way to reconcile responsibility and autonomy is to codify some of our responsibilities in advance of particular situations, rather than determining them through constant assessment and reassessment of particular situations.

Does this appeal to abstract rules mean that justice ignores our 'distinct individuality'? It is true that justice, in this context, does not require us to adjust our notion of 'what is reasonable to expect' to the particular needs of those around us. Our rights and obligations in these contexts are fixed in advance by abstract rules, not by context-sensitive assessments of the needs of those around us. But this should not be seen as evidence of an insensitivity to those particular needs. For the net result of this abstraction from particularity is to protect particularity more fully. The more our claims are dependent on context-sensitive calculations of everyone's particular desires, the more vulnerable our personal projects are to the shifting desires of others, and so the less we will be able to make long-term commitments. Meaningful autonomy requires predictability, and predictability requires some insulation from context-sensitivity.

This still leaves the possibility that some people will have strong desires that are frustrated by the application of abstract rules. But, as we have seen, the justice ethic assumes that competent adults are capable of adjusting their ends in the light of public standards. Assuming that the rules are publicly known, and confining our attention to competent adults for the moment, then the people who will suffer from the application of abstract rules are those who, through extravagance or carelessness, have formed desires which cannot be met within their rightfully allotted means. There may be such people in any particular situation, and their suffering may receive less notice in a society that appeals to abstract rules rather than context-sensitive assessments of particular needs. But this is their own responsibility, and it is unfair to ask others to make sacrifices to spare them from their irresponsibility.

The difficulty of staking out ground for personal autonomy within the care ethic is reminiscent of a similar problem within utilitarianism (ch. 2, s. 3*a* above). In both cases, the moral agent faces a seemingly 'unlimited responsibility' to 'act for the best in a causal framework formed to a considerable extent by the projects [of others]'. The agent's decisions become 'a function of all the satisfactions which he can affect from where he is: and this means that the projects of others, to an indeterminately great extent, determine his decision', leaving little room for the independent pursuit of his own desires and convictions (Williams 1973: 115).[17] This parallel should not be surprising, for while care theorists reject the utilitarian commitment to maximization, both theories tend to ground moral claims in subjective hurt and happiness, rather than objective unfairness. As a result, both theories interpret concern for others primarily as a matter of responding to their already given needs. But we can only protect fairness and autonomy if we view concern for others not solely in terms of responding to pre-existing preferences, but as something that should enter into the very formation of our preferences. Rather than taking into account people's specific aims in deciding on just distributions, people should take principles of justice into account in deciding on their aims and ambitions. As Rawls puts it, within the justice ethic, individuals are responsible for forming 'their aims and ambitions in the light of what they can reasonably expect'. Those who fail to do so may suffer the frustration of strong desires, but people know that 'the weight of their claims is not given by the strength or intensity of their wants and desires' (Rawls 1980: 545). Thus we shift from subjective hurt or happiness to objective unfairness as the basis for moral claims.

We can now see the kernel of truth underlying the two previous contrasts between care and justice. According to Tronto, justice emphasizes the learning of rules over the learning of moral sensitivities, and the applying of abstract principles over the making of context-sensitive assessments of particular needs. This debate over abstraction and context-sensitivity in our moral capacities and moral reasoning is often presented as distinct from the debate over rights and responsibilities as moral concepts. It is often viewed as an epistemological debate, as if justice theorists think that abstract principles are more 'objective' or 'rational', whereas care theorists reject notions of objectivity as epistemologically unsound (e.g. Jaggar 1983: 357; Young 1987: 60). I argued earlier that the entire contrast is overdrawn, since the sort of abstraction involved in justice reasoning does not necessarily compete with context-sensitivity (e.g. moral sensitivity is required to be a good juror). But we can now see that even where justice is less context-sensitive, the explanation is moral not epistemological. The reason why justice emphasizes learning and applying rules is that this is required for fairness and autonomy. If we are to have genuine autonomy, we must know in advance what our responsibilities are, and these assignments of responsibility must be insulated to some extent from context-sensitive assessments of particular desires. As a result, some subjective hurts must be discounted. And if some subjective hurts do not give rise to moral claims, then people need to know in advance which these are, so that they can adjust their aims accordingly. For both these reasons, we need rules which are more abstract and less context-sensitive.[18] So any differences that do exist concerning the importance of context-sensitivity in our moral capacities and moral reasoning are derived from more fundamental differences concerning the importance of fairness and personal responsibility as moral concepts. The first two contrasts are by-products of the third.

It is not surprising that this notion of responsibility for our ends is the fundamental difference between care and justice, given the pervasiveness of the public–domestic distinction in our thought. The assumption that we are responsible for our ends is plausible precisely to the extent that we exclude care for dependent others from the scope of justice. Rawls rejects the view that subjective desires are the standard of moral claims on the grounds that 'to argue this seems to presuppose that citizens' preferences are beyond their control as propensities or cravings which simply happen' (Rawls 1982b: 168–9). But this presupposition, of course, is quite true of many people. Rawls's rejection of subjective hurt as the basis for moral claims is plausible as long as we think only of (able-bodied and mentally competent) adults interacting in public life, while the sick, the helpless, and the young are kept safely out of view.[19] Rawls says that interactions between able-bodied adults are the 'fundamental case' of justice. But once we look beyond the public sphere, then the 'fundamental case' shifts, for as Willard

Gaylin puts it, 'All of us inevitably spend our lives evolving from an initial to a final stage of dependence. If we are fortunate enough to achieve power and relative independence along the way it is a transient and passing glory' (quoted in Zaretsky 1983: 193).

On the other hand, the assumption that subjective hurts give rise to moral claims is plausible to the extent that we generalize from the caring relationships involved in child-rearing. A baby is not at all responsible for its needs, and cannot be expected to attend to its parent's welfare: 'Children cannot reciprocate care equally, they require a degree of selflessness and attention that is specific to them.' But precisely for this reason, a parent's role 'may often require one to tolerate, accept, and try not to be hurt by, behaviour that would be quite intolerable or a cause for anger in most adult relationships. . . . To see female "virtues" or priorities as arising mainly out of relationships with children may lead to tendency to gloss over the ways in which resilience has become resignation and acceptance, attention has become chronic anxiety, and care and responsiveness chronic self-denial' (Grimshaw 1986: 251, 253).

Should we say that care applies to relations with dependants, while justice applies to relations with autonomous adults? One problem is that the distribution of care is itself an issue of justice. Justice theorists have tended to assume that some people (women) will 'naturally' desire to care for others, as part of their plan of life, so that the work of caring for dependants is not something that imposes moral obligations on all persons. But as Baier argues, we cannot view caring as simply one possible life-plan, rather than a moral constraint on every life-plan, for 'the encouragement of some to cultivate it while others do not could easily lead to exploitation of those who do. It obviously has suited some in most societies well enough that others take on the responsibilities of care (for the sick, the helpless, the young) leaving them free to pursue their own less altruistic goods.' And, of course, 'the long unnoticed moral proletariat were the domestic workers, mostly female' (Baier 1987*b*: 49–50). If we want to ensure that 'free affectivity' for some people does not 'rely on and exploit the usually unfree affectivity' of those who care for dependants, then our political theory 'cannot regard concern for new and future persons as an optional charity left for those with a taste for it. If the morality the theory endorses is to sustain itself, it must provide for its own continuers, not just take out a loan on a carefully encouraged maternal instinct' (Baier 1987*b*: 53–4; 1988: 328).

Moreover, as we have seen, the elimination of sexual inequality requires not only the redistribution of domestic labour, but also a breakdown in the sharp distinction between public and domestic. We need to find ways to integrate public life and parenting, for example, rather than segregating child-rearing into a separate sphere. But while this integration of public and domestic is required for sexual justice, it threatens to undermine the presuppositions of justice reasoning. For justice not only presupposes that we are autonomous adults, it seems to presuppose that we are adults *who are not care-givers for dependants*. Once people are responsible for attending to the (unpredictable) demands of dependants, they are no longer capable of guaranteeing their own predictability. Perhaps the whole picture of autonomy as the free pursuit of projects formed in the light of abstract standards presupposes that care for dependent others can be delegated to someone else, or to the state. It is interesting to note how little care theorists talk about the sort of autonomy that male theorists discuss at length—the setting of personal goals, the commitment to personal projects. According to Baier, the care perspective 'makes autonomy not even an ideal. . . . A certain sort of freedom is an ideal, namely freedom of thought and expression, but to "live one's own life in one's own way" is not likely to be among the aims of persons' (Baier 1987*a*: 46). Likewise Ruddick says that maternal thinking involves 'a fundamental metaphysical attitude' which she calls 'holding', 'governed by the priority of keeping over acquiring', in which preserving existing ties takes precedence over the pursuit of new ambitions (Ruddick 1980: 217; 1987: 242). On these views, the commitment to autonomy is not a commitment to staking out ground for the pursuit of personal projects, free from the shifting needs of particular others, but is rather a commitment to meeting those needs in a courageous and imaginative way, rather than a servile or deferential way. Any more expansive notion of autonomy can only come at the price of abandoning our responsibilities.[20]

Can we meet our responsibilities for dependent others without giving up the more robust picture of autonomy, and the notions of responsibility and justice that make it possible? It is too early to tell. Justice theorists have constructed impressive edifices by refining traditional notions of fairness and responsibility. However, by continuing the centuries-old neglect of the basic issues of child-rearing and care for dependants, these intellectual achievements are resting on unexamined and perilously shaky ground. Any adequate theory of sexual equality must confront these issues, and the traditional conceptions of discrimination and privacy that have hidden them from view.

NOTES

1. For introductions to these diverse strands of feminist thought, see Jaggar (1983); Nye (1988); Charvet (1982); Tong (1989).

2 In accepting this prevailing view that there is 'a Foundation in Nature' for the rule of the husband 'as the abler and the stronger' (Locke, in Okin 1979: 200), classical liberals created a serious contradiction for themselves. For they also argued that all humans are by nature equal, that nature provides no grounds for an inequality of rights. This, we have seen, was the point of their state-of-nature theories (ch. 3, s. 3 above). Why should the supposed fact that men are 'abler and stronger' justify unequal rights for women when, as Locke himself says, 'differences in excellence of parts or ability' do not justify unequal rights? One cannot both maintain equality amongst men as a class, on the grounds that differences in ability do not justify different rights, and also exclude women as a class, on the grounds that they are less able. If women are excluded on the grounds that the average woman is less able than the average man, then all men who are less able than the average man must also be excluded. As Okin puts it, 'If the basis of his individualism was to be firm, he needed to argue that individual women were equal with individual men, just as weaker men were with stronger ones' (Okin 1979: 199).

3. Perhaps communitarians cannot accept the dominance approach, since it supposes that people can put their roles in question in a way that communitarianism denies, or disapproves of (ch. 6, s. 3 above— on the tension between feminism and communitarianism, see Greschner 1989; Okin 1989b: 41–62; Friedman 1989). And since libertarians do not accept the limited equality of condition that underlies the difference approach, they are not about to accept the dominance approach. The other theories, however, seem capable of adopting the dominance approach. Mackinnon argues that the dominance approach is beyond the ken of liberalism because liberals aim at 'formal' or 'abstract' law that is 'transparent of substance'. I do not understand her contrast of 'form' and 'substance', or how it relates to traditional liberal principles of equality and freedom. Mackinnon often seems to equate liberalism with a particular stream of American constitutional interpretation.

4. See Rawls (1971: 128, 146). Rawls's account of the family is discussed in Okin (1987); Green (1986); English (1977); Kearns (1983). The 'Aristotelian hangover' of treating individuals as 'heads of households' remains common in political science (Stiehm 1983).

5. One explanation is that liberals maintained the same dismissive attitude towards the domestic realm as the ancients. Just as the ancients viewed the domestic sphere as something to be transcended in order to free men to participate in political life, so liberals viewed domestic life as something to be mastered in order to be free for social life. This seems to be part of the explanation for why Mill and Marx did not consider reproduction to be a realm of freedom and justice. They viewed the traditional woman's role as a merely 'natural' one, incapable of cultural development (cf. Jaggar 1983: ch. 4; Okin 1979: ch. 9).

6. Many feminists say that the domestic–public distinction arose with, or was reflected in, liberalism's separation of public and private spheres (e.g. Nicholson 1986: 201; Kennedy and Mendus 1987: 6–7; Coltheart 1986: 112). But this is historically inaccurate, for 'the assignment of public space to men and [domestic] space to women is continuous in Western history' (Eisenstein 1981: 22). Liberalism inherited, rather than created, this public–domestic distinction. It may be true that by emphasizing the distinction between public and private within civil society, liberals obscure the more fundamental distinction between public and domestic (Pateman 1987: 109). But if so, it is a pre-liberal distinction between male and female domains which is being obscured (Eisenstein 1981: 223; cf. Green 1986: 34; Nicholson 1986: 161).

   Why has this original liberal understanding of the private sphere been lost, so that 'to talk of an ideal of the private world within the context of contemporary American society is to talk about the family' (Elshtain 1981: 322; cf. Benn and Gaus 1983: 54)? Perhaps because people assume that 'public' and 'private' must mark a division in space. If so, then the most plausible location of private space is the family household. But the liberal public–private distinction is not a distinction between two physical areas, since society and polity are essentially conterminous. It is a distinction between two different aims and responsibilities. To act publicly is to accept responsibility for promoting the common good, defined in terms of the impartial concern for each person's interests. When acting privately, one is not required to act impartially, but is free to pursue one's own ends, consistent with the rights of others, and to join with others in the pursuit of shared ends. Both of these activities can occur anywhere in society. The fact that one goes out in public does not mean that one is responsible for acting impartially or obliged to account for one's actions. The fact that one is at home does not absolve one from respecting other people's rights.

7. There are feminist critics of the liberal state–society divide, even when it is distinguished from the patriarchal domestic–public divide. Pateman, for example, says that unlike republican critics who seek only to 'reinstate the political in public life', feminist critics 'insist that an alternative to the liberal conception must also encompass the relationship between public and domestic life' (Pateman 1987: 108). But she does not explain why feminists who reject the public–domestic distinction should also be concerned with the liberal public–private distinction. Her own comments suggest that we have no clear idea about how integrating politics into civil society would affect the opposition of public and domestic (Pateman 1987: 120). Frances Olsen gives a feminist critique of the state–society distinction, drawing on Marx's comments on alienation (Olsen 1983: 1561–4).

8. This Romantic view of privacy has become so integrated into modern liberalism that some people take it as the original liberal conception (e.g. Benn and Gaus 1983: 57–8). However, while the idea of retreating from society can be found in classic liberals (e.g. Locke's *Letter of Toleration*), it is primarily an adopted liberal position. Viewing privacy in terms of a retreat from all roles of civil society, far from being the original liberal position, means that 'the personal and private have been dissociated from virtually every institutional setting. The result is a dramatic collapse of the traditional liberal distinction between public and private as between government and society' (Rosenblum 1987: 66).

9. It is remarkable how policies that were justified on the grounds that the family was the man's private property are now justified on the grounds that men and women have an equal right to privacy (see e.g. Benn and Gaus 1983: 38). As Taub and Schneider note, 'The state's failure to regulate the domestic sphere is now often justified on the ground that the law should not interfere with emotional relationships involved in the family realm because it is too heavy-handed. . . . The importance of this concern, however, is undercut by the fact that the same result was previously justified by legal fictions, such as the woman's civil death on marriage' (Taub and Schneider 1982: 122).

10. Some liberal feminists have started to challenge the traditional family. The characterization of liberal feminism as concerned only with access to the public sphere 'has become increasingly problematic. Liberal feminists, like many others, have steadily focused their attention on women's personal lives' (Nicholson 1986: 22–3; cf. Wendell 1987). Paradoxically, when liberals endorse reforming the family, they are often accused of the 'devaluation of the private sphere' (Elshtain 1981: 243; cf. Nicholson 1986: 24). Jean Elshtain claims that the 'liberal

imperative' is 'to thoroughly politicize or publicize the private sphere' (Elshtain 1981: 248). By making child-rearing a public responsibility, liberalism would 'denude the private sphere of its central raison d'être and chief source of human emotion and value. Similarly, to externalize all housekeeping activities, to make them public activities, would vitiate the private realm further. "All persons would, so far as possible, be transformed into public persons, and the sundering of the forms of social life begun by industrialization would be carried to completion by the absorption of the private as completely as possible into the public". This is the completion of the liberal imperative' (Elshtain 1981: 248, quoting R. P. Wolff). For a discussion of recent feminist concerns about 'liberalizing the family' (e.g. the extension of contractual thinking to marriage and the family), see my 'Rethinking the Family', *Philosophy and Public Affairs*, vol. 20 (1991).

11. There is also some debate over the explanation for any gender difference in moral reasoning. Proposals range from sex-role socialization (Meyers 1987: 142–6), to our early infant experience of being mothered (Gilligan 1987: 20). There are also less gender-specific explanations. Powerless groups often learn empathy because they are dependent on others for protection, and 'as subordinates in a male-dominated society, [women] are required to develop psychological characteristics that please the dominant group and fulfill its needs' (Okin 1990: 154). For example, 'a woman who is dependent on a man may develop great skill in attending to and caring for him, in "reading" his behaviour and learning how to interpret his moods and gratify his desires before he needs to ask' (Grimshaw 1986: 252). This may explain why the members of oppressed classes or races exhibit some manifestations of a caring ethic (Tronto 1987: 649–51; Harding 1987: 307).

12. As Okin notes, Gilligan's studies do not confront the question of 'how women think when confronted with a moral dilemma involving a conflict between the needs or interests of family and close friends and the needs or interests of more distant "others"' (Okin 1990: 158). Ruddick's answer is that 'mothers can . . . come to realize that the good of their own children is entwined with the good of all children' (Ruddick 1984: 239; but cf. 1987: 250–1). But it is doubtful that one child's good is connected with the good of all children, however distant. And even where it is entwined, the problem is that the connection can be competitive rather than complementary. Their good can be entwined in such a way that resources spent on one child must be denied to another. If the mechanism for expanding the web of relationships is the realization that one's good is entwined with that of others, then it may be a very limited expansion. It seems hopelessly optimistic to say that attending to distant others imposes no costs on

ongoing attachments, or that 'inequity adversely affects both parties in an unequal relationship' (Gilligan 1982: 174). Only an explicit commitment to impartial concern, and not simply to preserving existing connections, could sustain the sort of generalization that Gilligan and Ruddick desire.

13. Iris Young offers a more general argument for the claim that 'the impartial point of view' denies differences: 'Impartial reason must judge from a point of view outside of the particular perspectives of persons involved in interaction, able to totalize these perspectives into a whole, or general will. . . . The impartial subject need acknowledge no other subjects whose perspective should be taken into account and with whom discussion might occur. . . . From this impartial point of view one need not consult with any other, because the impartial point of view already takes into account all possible perspectives' (Young 1987: 62). But, as we have seen, Rawlsian impartiality consists precisely in the requirement that we attend to all the possible viewpoints. It seems that Young is confusing the moral requirement of impartiality with an epistemological requirement of impersonality or objectivity: 'As a characteristic of reason, impartiality means something different from the pragmatic attitude of being fair, considering other people's needs and desires as well as one's own. Impartiality names a point of view of reason that stands apart from any interests or desires. Not to be partial means being able to see the whole, how all the particular perspectives and interests in a given moral situation relate to one another in a way that, because of its partiality, each perspective cannot see itself. The impartial moral reasoner thus stands outside of and above the situation about which he or she reasons, with no stake in it, or is supposed to adopt an attitude toward a situation as though he or she were outside and above it' (Young 1987: 60). However, one can accept the moral claims of the original position as a mechanism for considering other people's distinct interests without accepting the epistemological ideal of standing above the situation. (Conversely, rejecting that ideal of impersonality does not guarantee that people will attend to other people's interests.)

14. According to some commentators, the difficulty in reconciling the two ethics is not conceptual, but developmental. According to Gilligan, different components of moral development are rooted in fundamentally different childhood experiences—i.e. the child's experience of inequality/powerlessnesss gives rise to the search for independence and equality; whereas the experience of deep attachment and connection gives rise to compassion and love (Gilligan 1987: 20). If so, then differences in infant experiences of being parented may affect their ability to learn different components of morality (Flanagan and Jackson 1987: 629).

15. However, most justice theorists recognize Good Samaritan obligations which are unrelated to objective unfairness (ch. 2 n. 9 above).

16. Some commentators argue that Gilligan, by neglecting the issue of oppressive relationships, runs the danger of 'moral essentialism'. She 'separates the qualities of care and connection from their context of inequality and oppression and demands that they be considered in their own right, according to their intrinsic merit' (Houston 1988: 176). As Tronto notes, 'If the preservation of a web of relationships is the starting premise of an ethic of care, then there is little basis for critical reflection on whether those relationships are good, healthy, or worthy of preservation' (Tronto 1987: 660; cf. Wilson 1988: 17–18).

17. Hence it is quite misleading to say that Gilligan shares Williams's belief that impartiality is 'too demanding', or his hope that by emphasizing the importance of 'the personal point of view' we can free personal projects from the constraints of morality (*contra* Adler 1987: 226, 205; Kittay and Meyers 1987: 8). As Blum notes, personal concerns are seen by Williams 'as legitimate not so much from the standpoint of *morality*, but from the broader standpoint of practical reason. By contrast Gilligan argues . . . that care and responsibility within personal relationships constitute an important element of morality itself, genuinely distinct from impartiality. For Gilligan each person is embedded within a web of ongoing relationships, and morality importantly if not exclusively consists in attention to, understanding of, and emotional responsiveness toward the individuals with whom one stands in these relationships. . . . Nagel's and Williams's notions of the personal domain do not capture or encompass (though Nagel and Williams sometimes imply that they are meant to) the phenomenon of care and responsibility within personal relationships and do not explain why care and responsibility in relationships are distinctively moral phenomena' (Blum 1988: 473). Blum concludes that Gilligan's critique is 'importantly different' from Williams's critique of impartiality, but 'is not at odds' with it (Blum 1988: 473). But this still understates the problem, since Williams clearly wants to emphasize the *non-moral* value of personal projects, and wants to contain morality so as to protect these non-moral values. Gilligan wants to moralize the very attachments which Williams says have non-moral importance.

18. The argument for public standards is also relevant to democracy. The care ethic's claim that moral problems should be solved, not by appeal to public rules or principles, but through the exercise of moral sensitivities by the morally mature agent, has a strong similarity to conservative arguments that political leaders must not be held too accountable to the democratic process (e.g. Oakeshott 1984). Wise political leaders must be trusted, rather than scrutinized, for their

reasoning is often tacit, and impossible to present systematically. As with rules of justice, we may want political leaders to employ clear public standards of justification, not because they are more objective, but because they are more democratic. See Dietz (1985) for a critique of maternal thinking for ignoring political values like democracy.

19. Other theories, like those of Rawls and Dworkin, recognize that we have obligations towards dependent others (ch. 3, s. 4*b* above). But they write as if these obligations are a matter of ensuring that a fair share of resources is allocated to children and the infirm. They do not discuss our obligation to provide care for dependants.

20. For example, Leslie Wilson says that the reason why the 'ethical self of a person requires a certain sort of autonomy' is that it enables us 'to become the sort of person who can be genuinely one-caring'. Hence an autonomous person exercises her autonomy 'trying to determine ways in which one could become a better caring individual' (Wilson 1988: 21–2). Likewise, Ruddick says that the reason why attentive love requires 'realistic self-preservation', rather than 'chronic self-denial', is that we can become better caring individuals that way (Ruddick 1984: 238). This is some distance from the traditional picture of autonomy as the free pursuit of projects that matter to one for their own sake, and which occasionally compete for time and energy with one's moral obligations.

Bibliography

ACKERMAN, B. (1980). *Social Justice in the Liberal State*. Yale University Press, New Haven, Conn.

ADLER, J. (1987). 'Moral Development and the Personal Point of View', in Kittay and Meyers (1987).

ALEXANDER, L., and SCHWARZSCHILD, M. (1987). 'Liberalism, Neutrality, and Equality of Welfare vs. Equality of Resources', *Philosophy and Public Affairs*, 16/1: 85–110.

ALLEN, A. (1988). *Uneasy Access: Privacy for Women in a Free Society*. Rowman and Allanheld, Totowa, NJ.

ALLEN, D. (1973). 'The Utilitarianism of Marx and Engels', *American Philosophical Quarterly*, 10/3: 189–99.

ARENDT, H. (1959). *The Human Condition*. Anchor, New York.

ARNESON, R. (1981). 'What's Wrong with Exploitation?', *Ethics*, 91/2: 202–27.

—— (1985). 'Freedom and Desire', *Canadian Journal of Philosophy*, 15/3: 425–48.

—— (1987). 'Meaningful Work and Market Socialism', *Ethics*, 97/3: 517–45.

—— (1989). 'Equality and Equal Opportunity for Welfare', *Philosophical Studies*, 56: 77–93.

ARTHUR, J. (1987). 'Resource Acquisition and Harm', *Canadian Journal of Philosophy*, 17/2: 337–47.

BAIER, A. (1987a). 'Hume, the Women's Moral Theorist?', in Kittay and Meyers (1987).

—— (1987b). 'The Need for More than Justice', *Canadian Journal of Philosophy*, supplementary vol. 13: 14–56.

—— (1988). 'Pilgrim's Progress', *Canadian Journal of Philosophy*, 18/2: 315–30.

BAKER, C. (1985). 'Sandel on Rawls', *University of Pennsylvania Law Review*, 133/4: 895–928.

BARRY, B. (1973). *The Liberal Theory of Justice*. Oxford University Press, Oxford.

—— (1989). *Theories of Justice*. University of California Press, Berkeley, Calif.

BARRY, N. (1986). *On Classical Liberalism and Libertarianism*. Macmillan, London.

BEINER, R. (1983). *Political Judgment*. Methuen, London.

—— (1989). 'What's the Matter with Liberalism', in A. Hutchinson and L. Green (eds.), *Law and the Community*. Carswell, Toronto.

BELLAH, R., *et al.* (1985). *Habits of the Heart: Individualism and Commitment in American Life*. University of California Press, Berkeley, Calif.

BENHABIB, S. (1986). *Critique, Norm, and Utopia*. Columbia University Press, New York.

—— (1987). 'The Generalized and the Concrete Other: The Kohlberg–Gilligan Controversy and Feminist Theory', in S. Benhabib and D. Cornell (eds.), *Feminism as Critique*. University of Minnesota Press, Minneapolis, Minn.

BENN, S., and GAUS, G. (1983). *Public and Private in Social Life*. Croom Helm, Kent.

BERLIN, I. (1969). *Four Essays on Liberty*. Oxford University Press, London.

BERTRAM, C. (1988). 'A Critique of John Roemer's General Theory of Exploitation', *Political Studies*, 36/1: 123–30.

BLUM, L. (1988). 'Gilligan and Kohlberg: Implications for Moral Theory', *Ethics*, 98/3: 472–91.

BOGART, L. (1985). 'Lockean Provisos and State of Nature Theories', *Ethics*, 95/4: 828–36.

BRANDT, R. B. (1959). *Ethical Theory*. Prentice-Hall, Englewood Cliffs, NJ.

BRAVERMAN, H. (1974). *Labor and Monopoly Capital*. Monthly Review Press, New York.

BRENKERT, G. (1981). 'Marx's Critique of Utilitarianism', *Canadian Journal of Philosophy*, supplementary vol. 7: 193–220.

—— (1983). *Marx's Ethics of Freedom*. Routledge & Kegan Paul, London.

BRINK, D. (1986). 'Utilitarian Morality and the Personal Point of View', *Journal of Philosophy*, 83/8: 417–38.

BRITTAN, S. (1988). *A Restatement of Economic Liberalism*. Macmillan, London.

BROOME, J. (1990–1). 'Fairness', *Proceedings of the Aristotelian Society*, vol. 91: 87–102.

BROUGHTON, J. (1983). 'Women's Rationality and Men's Virtues', *Social Research*, 50/3: 597–642.

BROWN, A. (1986). *Modern Political Philosophy: Theories of the Just Society*. Penguin, Harmondsworth.

BUCHANAN, A. (1982). *Marx and Justice: The Radical Critique of Liberalism*. Methuen, London.

—— (1989). 'Assessing the Communitarian Critique of Liberalism', *Ethics*, 99/4: 852–82.

BUCHANAN, J. (1975). *The Limits of Liberty: Between Anarchy and Leviathan*. University of Chicago Press, Chicago, Ill.

CAMPBELL, T. (1983). *The Left and Rights: A Conceptual Analysis of the Idea of Socialist Rights*. Routledge & Kegan Paul, London.

—— (1988). *Justice*. Macmillan, Basingstoke.

CARENS, J. (1985). 'Compensatory Justice and Social Institutions', *Economics and Philosophy*, 1/1: 39–67.

—— (1986). 'Rights and Duties in an Egalitarian Society', *Political Theory*, 14/1: 31–49.

CAREY, G. (1984). *Freedom and Virtue: The Conservative/Libertarian Debate*. University Press of America, Lanham, Md.

CHARVET, J. (1982). *Feminism*. J. M. Dent and Sons, London.

CHRISTMAN, J. (1986). 'Can Ownership be Justified by Natural Rights?', *Philosophy and Public Affairs*, 15/2: 156–77.

CLARK, B., and GINTIS, H. (1978). 'Rawlsian Justice and Economic Systems', *Philosophy and Public Affairs*, 7/4: 302–25.

COHEN, G. A. (1978). *Karl Marx's Theory of History: A Defence*. Princeton University Press, Princeton, NJ.

—— (1979). 'Capitalism, Freedom and the Proletariat', in A. Ryan (ed.), *The Idea of Freedom*. Oxford University Press, Oxford.

—— (1986a). 'Self-Ownership, World-Ownership and Equality', in F. Lucash (ed.), *Justice and Equality Here and Now*. Cornell University Press, Ithaca, NY.

—— (1986b). 'Self-Ownership, World-Ownership and Equality: Part 2', *Social Philosophy and Policy*, 3/2: 77–96.

—— (1988). *History, Labour, and Freedom: Themes from Marx*. Oxford University Press, Oxford.

—— (1989). 'On the Currency of Egalitarian Justice', *Ethics*, 99/4: 906–44.

—— (1990a). 'Marxism and Contemporary Political Philosophy, or: Why Nozick Exercises some Marxists more than he does any Egalitarian Liberals', *Canadian Journal of Philosophy*, supplementary vol. 16: 363–87.

—— (1990b). 'Self-Ownership, Communism and Equality', *Proceedings of the Aristotelian Society*, supplementary vol. 64.

COLTHEART, D. (1986). 'Desire, Consent and Liberal Theory', in C. Pateman and E. Gross (eds.), *Feminist Challenges: Social and Political Theory*. Northeastern University Press, Boston, Mass.

CONNOLLY, W. (1984). 'The Dilemma of Legitimacy', in W. Connolly (ed.), *Legitimacy and the State*. Blackwell, Oxford.

COPP, D. (1990). 'Contractarianism and Moral Skepticism', in P. Vallentyne (ed.), *Contractarianism and Rational Choice: Essays on Gauthier*. Cambridge University Press, New York.

CRAGG, W. (1986). 'Two Concepts of Community or Moral Theory and Canadian Culture', *Dialogue*, 25/1: 31–52.

CROCKER, L. (1977). 'Equality, Solidarity, and Rawls' Maximin', *Philosophy and Public Affairs*, 6/3: 262–6.

CROWLEY, B. (1987). *The Self, the Individual and the Community: Liberalism in the Political Thought of F. A. Hayek and Sidney and Beatrice Webb*. Oxford University Press, Oxford.

DANIELS, N. (1975). 'Equal Liberty and Unequal Worth of Liberty', in N. Daniels (ed.), *Reading Rawls*. Basic Books, New York.

—— (1985). *Just Health Care*. Cambridge University Press, Cambridge.

DICK, J. (1975). 'How to Justify a Distribution of Earnings', *Philosophy and Public Affairs*, 4/3: 248–72.

DIETZ, M. (1985). 'Citizenship with a Feminist Face: The Problem with Maternal Thinking', *Political Theory*, 13/1: 19–37.

DIGGS, B. J. (1981). 'A Contractarian View of Respect for Persons', *American Philosophical Quarterly*, 18/4: 273–83.

DiQUATTRO, A. (1983). 'Rawls and Left Criticism', *Political Theory*, 11/1: 53–78.

DOPPELT, G. (1981). 'Rawls' System of Justice: A Critique from the Left', *Nous*, 15/3: 259–307.

DWORKIN, R. (1977). *Taking Rights Seriously*. Duckworth, London.

—— (1978). 'Liberalism', in S. Hampshire (ed.), *Public and Private Morality*. Cambridge University Press, Cambridge.

—— (1981). 'What is Equality? Part I: Equality of Welfare; Part II: Equality of Resources', *Philosophy and Public Affairs*, 10/3–4: 185–246, 283–345.

—— (1983). 'In Defense of Equality', *Social Philosophy and Policy*, 1/1: 24–40.

—— (1985). *A Matter of Principle*. Harvard University Press, London.

—— (1986). *Law's Empire*. Harvard University Press, Cambridge, Mass.

—— (1987). 'What is Equality? Part III: The Place of Liberty'. *Iowa Law Review*, 73/1: 1–54.

—— (1988). 'What is Equality? Part 4: Political Equality', *University of San Francisco Law Review*, 22/1: 1–30.

—— (1989). 'Liberal Community', *California Law Review*, 77/3: 479–504.

EHRENREICH, B., and ENGLISH, J. (1973). *Witches, Midwives, and Nurses: A History of Women Healers*. Feminist Press, Old Westbury.

EICHBAUM, J. (1979). 'Towards an Autonomy-Based Theory of Constitutional Privacy: Beyond the Ideology of Familial Privacy', *Harvard Civil Rights–Civil Liberties Law Review*, 14/2: 361–84.

EISENSTEIN, Z. (1981). *The Radical Future of Liberal Feminism*. Longman, New York.

—— (1984). *Feminism and Sexual Equality: Crisis in Liberal America.* Monthly Review Press, New York.

ELSHTAIN, J. (1981). *Public Man, Private Women: Women in Social and Political Thought.* Princeton University Press, Princeton, NJ.

ELSTER, J. (1982). 'Roemer vs. Roemer', *Politics and Society*, 11/3: 363–73.

—— (1983). 'Exploitation, Freedom, and Justice', in J. R. Pennock and J. W. Chapman (eds.), *Marxism: Nomos 26.* New York University Press, New York.

—— (1985). *Making Sense of Marx.* Cambridge University Press, Cambridge.

—— (1986). 'Self-Realization in Work and Politics: The Marxist Conception of the Good Life', *Social Philosophy and Policy*, 3/2: 97–126.

ENGELS, F. (1972). *The Origin of the Family, Private Property, and the State.* International Publishers, New York.

ENGLISH, J. (1977). 'Justice between Generations', *Philosophical Studies*, 31/2: 91–104.

EVANS, S. (1979). *Personal Politics: The Roots of Women's Liberation in the Civil Rights Movement and the New Left.* Knopf, New York.

EXDELL, J. (1977). 'Distributive Justice: Nozick on Property Rights', *Ethics*, 87/2: 142–9.

FEINBERG, J. (1980). *Rights, Justice, and the Bounds of Liberty.* Princeton University Press, Princeton, NJ.

—— (1988). *Harmless Wrongdoing.* Vol. iv of *The Moral Limits of the Criminal Law.* Oxford University Press, Oxford.

FLANAGAN, O., and ADLER, J. (1983). 'Impartiality and Particularity', *Social Research*, 50/3: 576–96.

—— and JACKSON, K. (1987). 'Justice, Care, and Gender: The Kolhberg–Gilligan Debate Revisited', *Ethics*, 97/3: 622–37.

FLEW, A. (1979). *A Dictionary of Philosophy.* Pan Books, London.

—— (1989). *Equality in Liberty and Justice.* Routledge, London.

FREY, R. (1984). *Utility and Rights.* University of Minnesota Press, Minneapolis, Minn.

FRIED, C. (1978). *Right and Wrong.* Harvard University Press, Cambridge, Mass.

—— (1983). 'Distributive Justice', *Social Philosophy and Policy*, 1/1: 45–59.

FRIEDMAN, M. (1987*a*). 'Beyond Caring: The De-moralization of Gender', *Canadian Journal of Philosophy*, supplementary vol. 13: 87–110.

—— (1987*b*). 'Care and Context in Moral Reasoning', in Kittay and Meyers (1987).

—— (1989).'Feminism and Modern Friendship: Dislocating the Community', *Ethics*, 99/2: 275–90.

FRYE, M. (1983). *The Politics of Reality: Essays in Feminist Theory.* Crossing Press, Trumansburg.

FUNK, N. (1988). 'Habermas and the Social Goods', *Social Text,* 18: 19–37.

GALSTON, W. (1980). *Justice and the Human Good.* University of Chicago Press, Chicago, Ill.

—— (1986). 'Equality of Opportunity and Liberal Theory', in F. Lucash (ed.), *Justice and Equality Here and Now.* Cornell University Press, Ithaca, NY.

GAUTHIER, D. (1986). *Morals by Agreement.* Oxford University Press, Oxford.

GERAS, N. (1989). 'The Controversy about Marx and Justice', in A. Callinicos (ed.), *Marxist Theory.* Oxford University Press, Oxford.

GIBBARD, A. (1985). 'What's Morally Special about Free Exchange?', *Social Philosophy and Policy,* 2/2: 20–8.

GILLIGAN, C. (1982). *In a Different Voice: Psychological Theory and Women's Development.* Harvard University Press, Cambridge, Mass.

—— (1986). 'Remapping the Moral Domain', in T. Heller, M. Sosna, and D. Wellbury (eds.), *Reconstructing Individualism: Autonomy, Individuality, and the Self in Western Thought.* Stanford University Press, Stanford, Calif.

—— (1987). 'Moral Orientation and Moral Development', in Kittay and Meyers (1987).

GOODIN, R. (1982). *Political Theory and Public Policy.* University of Chicago Press, Chicago, Ill.

—— (1988). *Reasons for Welfare.* Princeton University Press, Princeton, NJ.

GORDON, S. (1980). *Welfare, Justice, and Freedom.* Columbia University Press, New York.

GOUGH, J. W. (1957). *The Social Contract.* 2nd edn., Oxford University Press, London.

GOULD, C. (1978). *Marx's Social Ontology.* MIT Press, Cambridge, Mass.

GRAY, J. (1986a). *Liberalism.* University of Minnesota Press, Minneapolis, Minn.

—— (1986b). 'Marxian Freedom, Individual Liberty, and the End of Alienation', *Social Philosophy and Policy,* 3/2: 160–87.

—— (1989). *Liberalisms: Essays in Political Philosophy.* Routledge, London.

GREEN, K. (1986). 'Rawls, Women and the Priority of Liberty', *Australasian Journal of Philosophy,* supplement to vol. 64: 26–36.

GRESCHNER, D. (1989). 'Feminist Concerns with the New Communitarians', in A. Hutchinson and L. Green (eds.), *Law and the Community.* Carswell, Toronto.

GREY, T. (1980). 'Eros, Civilization, and the Burger Court', *Law and Contemporary Problems*, 43/3: 83–100.

GRICE, G. (1967). *The Grounds of Moral Judgement*. Cambridge University Press, Cambridge.

GRIFFIN, J. (1986). *Well-Being: Its Meaning, Measurement, and Moral Importance*. Oxford University Press, Oxford.

GRIMSHAW, J. (1986). *Philosophy and Feminist Thinking*. University of Minnesota Press, Minneapolis, Minn.

GROSS, E. (1986). 'What is Feminist Theory?', in C. Pateman and E. Gross (eds.), *Feminist Challenges, Social and Political Theory*. Northeastern University Press, Boston, Mass.

GUTMANN, A. (1980). *Liberal Equality*. Cambridge University Press, Cambridge.

—— (1985). 'Communitarian Critics of Liberalism', *Philosophy and Public Affairs*, 14/3: 308–22.

HABERMAS, J. (1979). *Communication and the Evolution of Society*, trans. T. McCarthy. Beacon, Boston, Mass.

—— (1985). 'Questions and Counterquestions', in R. Bernstein (ed.), *Habermas and Modernity*. MIT Press, Cambridge, Mass.

HAMPTON, J. (1980). 'Contracts and Choices: Does Rawls Have a Social Contract Theory?', *Journal of Philosophy*, 77/6: 315–38.

—— (1986). *Hobbes and the Social Contract Tradition*. Cambridge University Press, Cambridge.

HARDING, S. (1982). 'Is Gender a Variable in Conceptions of Rationality? A Survey of Issues', *Dialectica*, 36/2: 225–42.

—— (1987). 'The Curious Coincidence of Feminine and African Moralities', in Kittay and Meyers (1987).

HARE, R. M. (1963). *Freedom and Reason*. Oxford University Press, London.

—— (1971). *Essays on Philosophical Method*. Macmillan, London.

—— (1975). 'Rawls' Theory of Justice', in N. Daniels (ed.), *Reading Rawls*. Basic Books, New York.

—— (1978). 'Justice and Equality', in J. Arthur and W. Shaw (eds.), *Justice and Economic Distribution*. Prentice-Hall, Englewood Cliffs, NJ.

—— (1982). 'Ethical Theory and Utilitarianism', in A. Sen and B. Williams (eds.), *Utilitarianism and Beyond*. Cambridge University Press, Cambridge.

—— (1984). 'Rights, Utility, and Universalization: Reply to J. L. Mackie', in Frey (1984).

HARMAN, G. (1983). 'Human Flourishing, Ethics, and Liberty', *Philosophy and Public Affairs*, 12/4: 307–22.

HARSANYI, J. (1976). *Essays on Ethics, Social Behavior and Scientific Explanation*. Reidel, Dordrecht.

HARSANYI, J. (1977). *Rational Behavior and Bargaining Equilibrium in Games and Social Situations*. Cambridge University Press, Cambridge.

—— (1985). 'Rule Utilitarianism, Equality, and Justice', *Social Philosophy and Policy*, 2/2: 115–27.

HART, H. L. A. (1975). 'Rawls on Liberty and its Priority', in N. Daniels (ed.), *Reading Rawls*. Basic Books, New York.

—— (1979). 'Between Utility and Rights', in A. Ryan (ed.), *The Idea of Freedom*. Oxford University Press, Oxford.

HASLETT, D. (1987). *Equal Consideration: A Theory of Moral Justification*. University of Delaware, Newark, NJ.

HAYEK, F. A. (1960). *The Constitution of Liberty*. Routledge & Kegan Paul, London.

HELD, V. (1987). 'Feminism and Moral Theory', in Kittay and Meyers (1987).

HERZOG, D. (1986). 'Some Questions for Republicans', *Political Theory*, 14/3: 473–93.

HIRSCH, H. (1986). 'The Threnody of Liberalism: Constitutional Liberty and the Renewal of Community', *Political Theory*, 14/3: 423–49.

HOLMES, S. (1989). 'The Permanent Structure of Antiliberal Thought', in N. Rosenblum (ed.), *Liberalism and the Moral Life*. Harvard University Press, Cambridge, Mass.

HOLMSTROM, N. (1977). 'Exploitation', *Canadian Journal of Philosophy*, 7/2: 353–69.

HOSPERS, J. (1961). *Human Conduct: An Introduction to the Problem of Ethics*. Harcourt, Brace and World, New York.

HOUSTON, B. (1988). 'Gilligan and the Politics of a Distinctive Women's Morality', in L. Code, S. Mullett, and C. Overall (eds.), *Feminist Perspectives: Philosophical Essays on Method and Morals*. University of Toronto Press, Toronto.

JAGGAR, A. (1983). *Feminist Politics and Human Nature*. Rowman and Allanheld, Totowa, NJ.

JONES, P. (1982). 'Freedom and the Redistribution of Resources', *Journal of Social Policy*, 11/2: 217–38.

KEARNS, D. (1983). 'A Theory of Justice—and Love: Rawls on the Family', *Politics*, 18/2: 36–42.

KEAT, R. (1982). 'Liberal Rights and Socialism', in K. Graham (ed.), *Contemporary Political Philosophy: Radical Studies*. Cambridge University Press, Cambridge.

KENNEDY, E., and MENDUS, S. (1987). *Women in Western Political Philosophy*. Wheatsheaf Books, Brighton.

KERNOHAN, A. (1988). 'Capitalism and Self-Ownership', *Social Philosophy and Policy*, 6/1: 60–76.

KITTAY, E., and MEYERS, D. (1987). *Women and Moral Theory*. Rowman and Littlefield, Savage, Md.

KOHLBERG, L. (1984). *Essays on Moral Development*, ii. Harper and Row, San Francisco, Calif.

KROUSE, R., and MCPHERSON, M. (1988). 'Capitalism, "Property-Owning Democracy", and the Welfare State', in A. Gutmann (ed.), *Democracy and the Welfare State*. Princeton University Press, Princeton, NJ.

KYMLICKA, W. (1988a). 'Liberalism and Communitarianism', *Canadian Journal of Philosophy*, 18/2: 181–203.

—— (1988b). 'Rawls on Teleology and Deontology', *Philosophy and Public Affairs*, 17/3: 173–90.

—— (1989a). *Liberalism, Community, and Culture*. Oxford University Press, Oxford.

—— (1989b). 'Liberal Individualism and Liberal Neutrality', *Ethics*, 99/4: 883–905.

LADENSON, R. (1983). *A Philosophy of Free Expression and its Constitutional Applications*. Rowman and Littlefield, Totowa, NJ.

LARMORE, C. (1987). *Patterns of Moral Complexity*. Cambridge University Press, Cambridge.

LESSNOFF, M. (1986). *Social Contract*. Macmillan, London.

LEVINE, A. (1988). 'Capitalist Persons', *Social Philosophy and Policy*, 6/1: 39–59.

—— (1989). 'What is a Marxist Today?', *Canadian Journal of Philosophy*, supplementary vol. 15: 29–58.

LINDBLOM, C. (1977). *Politics and Markets*. Basic Books, New York.

LOEVINSOHN, E. (1977). 'Liberty and the Redistribution of Property', *Philosophy and Public Affairs*, 6/3: 226–39.

LOMASKY, L. (1987). *Persons, Rights, and the Moral Community*. Oxford University Press, Oxford.

LUKES, S. (1985). *Marxism and Morality*. Oxford University Press, Oxford.

LYONS, D. (1965). *Forms and Limits of Utilitarianism*. Oxford University Press, London.

—— (1981). 'The New Indian Claims and Original Rights to Land', in Paul (1981).

MACCALLUM, G. (1967). 'Negative and Positive Freedom', *Philosophical Review*, 76/3: 312–34.

MACEDO, S. (1988). 'Capitalism, Citizenship and Community', *Social Philosophy and Policy*, 6/1: 113–39.

MACINTYRE, A. (1981). *After Virtue: A Study in Moral Theory*. Duckworth, London.

MACKIE, J. (1984). 'Rights, Utility, and Universalization', in Frey (1984).

MACKINNON, C. (1987). *Feminism Unmodified: Discourses on Life and Law*. Harvard University Press, Cambridge, Mass.

MACPHERSON. C. B. (1973). *Democratic Theory: Essays in Retrieval*. Oxford University Press, Oxford.

MAPEL, D. (1989). *Social Justice Reconsidered*. University of Illinois Press, Urbana, Ill.

MARTIN, R. (1985). *Rawls and Rights*. University Press of Kansas, Lawrence, Kan.

MARX, K. (1973). *Grundrisse*, ed. M. Nicolaus. Penguin, Harmondsworth.

—— (1977*a*). *Economic and Philosophic Manuscripts of 1844*. Lawrence and Wishart, London.

—— (1977*b*). *Karl Marx: Selected Writings*, ed. D. McLellan. Oxford University Press, Oxford.

—— (1977*c*). *Capital: A Critique of Political Economy*, vol. i. Penguin, Harmondsworth.

—— (1981). *Capital: A Critique of Political Economy*, vol. iii. Penguin, Harmondsworth.

—— and ENGELS, F. (1968). *Marx/Engels: Selected Works in One Volume*. Lawrence and Wishart, London.

—— —— (1970). *The German Ideology*, ed. C. Arthur. Lawrence and Wishart, London.

MEYERS, D. (1987). 'The Socialized Individual and Individual Autonomy', in Kittay and Meyers (1987).

MICHELMAN, F. (1975). 'Constitutional Welfare Rights and *A Theory of Justice*', in N. Daniels (ed.), *Reading Rawls*. Basic Books, New York.

MIDGLEY, M. (1978). *Beast and Man: The Roots of Human Nature*. New American Library, New York.

MILL, J. S. (1962). *Mill on Bentham and Coleridge*, ed. F. Leavis. Chatto and Windus, London.

—— (1965). *Principles of Political Economy*, in *Collected Works*, iii. University of Toronto Press, Toronto.

—— (1967). 'Chapters on Socialism', in *Collected Works*, v. University of Toronto Press, Toronto.

—— (1968). *Utilitarianism, Liberty, Representative Government*, ed. A. D. Lindsay. J. M. Dent and Sons, London.

—— (1974). *On Liberty*, ed. G. Himmelfarb. Penguin, Harmondsworth.

—— (1989). 'In What Sense must Socialism be Communitarian?', *Social Philosophy and Policy*, 6/2: 51–73.

—— and MILL, H. T. (1970). *Essays on Sex Equality*, ed. A. Rossi. University of Chicago Press, Chicago, Ill.

MILLER, D. (1976). *Social Justice*. Oxford University Press, Oxford.

—— (1989). 'In what Sense must Socialism be Communitarian?', *Social Philosophy and Policy*, 6/2: 51–73.

MILLER, R. (1984). *Analyzing Marx*. Princeton University Press, Princeton, NJ.

MOORE, G. E. (1912). *Ethics*. Oxford University Press, London.

MORRIS, C. (1988). 'The Relation between Self-Interest and Justice in Contractarian Ethics', *Social Philosophy and Policy*, 5/2: 119–53.

MURPHY, J. (1973). 'Marxism and Retribution', *Philosophy and Public Affairs*, 2/3: 214–41.

NAGEL, T. (1979). *Mortal Questions*, Cambridge University Press, Cambridge.

—— (1980). 'The Limits of Objectivity', in S. McMurrin (ed.), *The Tanner Lectures on Human Values*, i. University of Utah Press, Salt Lake City, Utah.

—— (1981). 'Libertarianism without Foundations', in Paul (1981).

—— (1986). *The View from Nowhere*. Oxford University Press, New York.

NARVESON, J. (1983). 'On Dworkinian Equality', *Social Philosophy and Policy*, 1/1: 1–23.

—— (1988). *The Libertarian Idea*. Temple University Press, Philadelphia, Pa.

NICHOLSON, L. (1986). *Gender and History: The Limits of Social Theory in the Age of the Family*. Columbia University Press, New York.

NIELSEN, K. (1978). 'Class and Justice', in J. Arthur and W. Shaw (eds.), *Justice and Economic Distribution*. Prentice-Hall, Englewood Cliffs, NJ.

—— (1985). *Equality and Liberty: A Defense of Radical Egalitarianism*. Rowman and Allanheld, Totowa, NJ

—— (1987). 'Rejecting Egalitarianism: On Miller's Nonegalitarian Marx', *Political Theory*, 15/3: 411–23.

—— (1989). *Marxism and the Moral Point of View*. Westview Press, Boulder, Colo.

NODDINGS, N. (1984). *Caring: A Feminine Approach to Ethics and Moral Education*. University of California Press, Berkeley, Calif.

NOVE, A. (1983). *The Economics of Feasible Socialism*. George Allen and Unwin, London.

NOZICK, R. (1974). *Anarchy, State, and Utopia*. Basic Books, New York.

—— (1981). *Philosophical Explanations*. Harvard University Press, Cambridge, Mass.

NUNNER-WINKLER, G. (1984). 'Two Moralities?', in W. Kurtines and J. Gewirtz (eds.), *Morality, Moral Behavior and Moral Development*. John Wiley, New York.

NYE, A. (1988). *Feminist Theory and the Philosophies of Man*. Croom Helm, London.

OAKESHOTT, M. (1984). 'Political Education', in M. Sandel (ed.), *Liberalism and its Critics*. Blackwell, Oxford.

O'BRIEN, M. (1981). *The Politics of Reproduction*. Routledge & Kegan Paul, London.

OKIN, S. (1979). *Women in Western Political Thought*. Princeton University Press, Princeton, NJ.

—— (1981). 'Women and the Making of the Sentimental Family', *Philosophy and Public Affairs*, 11/1: 65–88.

OKIN, S. (1987). 'Justice and Gender', *Philosophy and Public Affairs*, 16/1: 42–72.

—— (1989*a*). 'Reason and Feeling in Thinking About Justice', *Ethics*, 99/2: 229–49.

—— (1989*b*). *Justice, Gender, and the Family*. Basic Books, New York.

—— (1990). 'Thinking like a Woman', in D. Rhode (ed.), *Theoretical Perspectives on Sexual Difference*. Yale University Press, New Haven, Conn.

OLSEN, F. (1983). 'The Family and the Market: A Study of Ideology and Legal Reform', *Harvard Law Review*, 96/7: 1497–578.

O'NEILL, O. (1980). 'The Most Extensive Liberty', *Proceedings of the Aristotelian Society*, 80: 45–59.

PAREKH, B. (1982). *Contemporary Political Thinkers*. Martin Robertson, Oxford.

PARFIT, D. (1984). *Reasons and Persons*. Oxford University Press, Oxford.

PATEMAN, C. (1975). 'Sublimation and Reification: Locke, Wolin and the Liberal Democratic Conception of the Political', *Politics and Society*, 5/4: 441–67.

—— (1980). ' "The Disorder of Women": Women, Love, and the Sense of Justice', *Ethics*, 91/1: 20–34.

—— (1987). 'Feminist Critiques of the Public/Private Dichotomy', in A. Phillips (ed.), *Feminism and Equality*. Blackwell, Oxford.

PAUL, J. (1981). *Reading Nozick*. Rowman and Littlefield, Totowa, NJ.

PETTIT, P. (1980). *Judging Justice: An Introduction to Contemporary Political Philosophy*. Routledge & Kegan Paul, London.

POGGE, T. (1989). *Realizing Rawls*. Cornell University Press, Ithaca, NY.

RADCLIFFE RICHARDS, J. (1980). *The Sceptical Feminist: A Philosophical Enquiry*. Routledge & Kegan Paul, London.

RAILTON, P. (1984). 'Alienation, Consequentialism, and the Demands of Morality', *Philosophy and Public Affairs*, 13/2: 134–71.

RAPHAEL, D. D. (1970). *Problems of Political Philosophy*. Pall Mall, London.

—— (1981). *Moral Philosophy*. Oxford University Press, Oxford.

RAWLS, J. (1971). *A Theory of Justice*. Oxford University Press, London.

—— (1974). 'Reply to Alexander & Musgrave', *Quarterly Journal of Economics*, 88/4: 633–55.

—— (1975). 'Fairness to Goodness', *Philosophical Review*, 84: 536–54.

—— (1978). 'The Basic Structure as Subject', in A. Goldman and J. Kim (eds.), *Values and Morals*. Reidel, Dordrecht.

—— (1979). 'A Well-Ordered Society', in P. Laslett and J. Fishkin (eds.), *Philosophy, Politics, and Society*. Fifth series. Yale University Press, New Haven, Conn.

—— (1980). 'Kantian Constructivism in Moral Theory', *Journal of Philosophy*, 77/9: 515–72.

—— (1982*a*). 'The Basic Liberties and their Priority', in S. McMurrin (ed.), *The Tanner Lectures on Human Values*, iii. University of Utah Press, Salt Lake City, Utah.

—— (1982*b*). 'Social Unity and Primary Goods', in A. Sen and B. Williams (eds.), *Utilitarianism and Beyond*. Cambridge University Press, Cambridge.

—— (1985). 'Justice as Fairness: Political not Metaphysical', *Philosophy and Public Affairs*, 14/3: 223–51.

—— (1988). 'The Priority of Right and Ideas of the Good', *Philosophy and Public Affairs*, 17/4: 251–76.

RAZ, J. (1986). *The Morality of Freedom*. Oxford University Press, Oxford.

REIMAN, J. (1981). 'The Possibility of a Marxian Theory of Justice', *Canadian Journal of Philosophy*, supplementary vol. 7: 307–22.

—— (1983). 'The Labor Theory of the Difference Principle', *Philosophy and Public Affairs*, 12/2: 133–59.

—— (1987). 'Exploitation, Force, and the Moral Assessment of Capitalism: Thoughts on Roemer and Cohen', *Philosophy and Public Affairs*, 16/1: 3–41.

—— (1989). 'An Alternative to "Distributive" Marxism: Further Thoughts on Roemer, Cohen, and Exploitation', *Canadian Journal of Philosophy*, supplementary vol. 15: 299–331.

RESCHER, N. (1966). *Distributive Justice: A Constructive Critique of the Utilitarian Theory of Distribution*. Bobbs-Merrill, Indianapolis, Ind.

RICH, A. (1979). *On Lies, Secrets and Silence: Selected Prose, 1966–1978*. Norton, New York.

ROEMER, J. (1982*a*). *A General Theory of Exploitation and Class*. Harvard University Press, Cambridge, Mass.

—— (1982*b*). 'Property Relations vs. Surplus Value in Marxian Exploitation', *Philosophy and Public Affairs*, 11/4: 281–313.

—— (1982*c*). 'New Directions in the Marxian Theory of Exploitation and Class', *Politics and Society*, 11/3: 253–87.

—— (1985*a*). 'Equality of Talent', *Economics and Philosophy*, 1/2: 151–87.

—— (1985*b*). 'Should Marxists Be Interested in Exploitation?', *Philosophy and Public Affairs*, 14/1: 30–65.

—— (1986). 'The Mismarriage of Bargaining Theory and Distributive Justice', *Ethics*, 97/1: 88–110.

—— (1988). *Free to Lose: An Introduction to Marxist Economic Philosophy*. Harvard University Press, Cambridge, Mass.

—— (1989). 'Second Thoughts on Property Relations and Exploitation', *Canadian Journal of Philosophy*, supplementary vol. 15: 257–66.

RORTY, R. (1985). 'Postmodernist Bourgeois Liberalism', in R. Hollinger (ed.), *Hermeneutics and Praxis*. University of Notre Dame Press, Notre Dame, Ind.

ROSENBLUM, N. (1987). *Another Liberalism: Romanticism and the Reconstruction of Liberal Thought*. Harvard University Press, Cambridge, Mass.

ROSS, W. D. (1930). *The Right and the Good*. Oxford University Press, London.

ROTHBARD, M. (1982). *The Ethics of Liberty*. Humanities Press, Atlantic Highlands, NJ.

RUDDICK, S. (1984a). 'Maternal Thinking', in J. Trebilcot (ed.), *Mothering: Essays in Feminist Theory*. Rowman and Allanheld, Totowa, NJ.

—— (1984b). 'Preservative Love and Military Destruction', in J. Trebilcot (ed.), *Mothering: Essays in Feminist Theory*. Rowman and Allanheld, Totowa, NJ.

—— (1987). 'Remarks on the Sexual Politics of Reason', in Kittay and Meyers (1987).

SANDEL, M. (1982). *Liberalism and the Limits of Justice*. Cambridge University Press, Cambridge.

—— (1984a). 'The Procedural Republic and the Unencumbered Self', *Political Theory*, 12/1: 81–96.

—— (1984b). 'Morality and the Liberal Ideal', *New Republic*, 7 May 1984, 190: 15–17.

—— (1989). 'Moral Argument and Liberal Toleration: Abortion and Homosexuality', *California Law Review*, 77/3: 521–38.

SARTORIUS, R. (1969). 'Utilitarianism and Obligation', *Journal of Philosophy*, 66/3: 67–81.

SCANLON, T. (1982). 'Contractualism and Utilitarianism', in A. Sen and B. Williams (eds.), *Utilitarianism and Beyond*. Cambridge University Press, Cambridge.

—— (1983). 'Freedom of Expression and Categories of Expression', in D. Copp and S. Wendell (eds.), *Pornography and Censorship*. Prometheus, Buffalo, NY.

—— (1988). 'The Significance of Choice', in S. McMurrin (ed.), *The Tanner Lectures on Human Values*, viii. University of Utah Press, Salt Lake City, Utah.

SCHWARTZ, A. (1982). 'Meaningful Work', *Ethics*, 92/4: 634–46.

SCHWARTZ, N. (1979). 'Distinction between Public and Private Life: Marx on the *zoon politikon*', *Political Theory*, 7/2: 245–66.

SCHWEICKART, D. (1978). 'Should Rawls be a Socialist?', *Social Theory and Practice*, 5/1: 1–27.

SEN, A. (1980). 'Equality of What?', in S. McMurrin (ed.), *The Tanner Lectures on Human Values*, i. University of Utah Press, Salt Lake City, Utah.

—— (1985). 'Rights and Capabilities', in T. Honderich (ed.), *Morality and Objectivity*. Routledge & Kegan Paul, London.

—— (1990). 'Justice: Means versus Freedom', *Philosophy and Public Affairs*, 19/2: 111–21.

SHER, G. (1975). 'Justifying Reverse Discrimination in Employment', *Philosophy and Public Affairs*, 4/2: 159–70.

—— (1987). 'Other Voices, Other Rooms? Women's Psychology and Moral Theory', in Kittay and Meyers (1987).

SINGER, P. (1979). *Practical Ethics*. Cambridge University Press, Cambridge.

SMART, J. J. C. (1973). 'An Outline of a System of Utilitarian Ethics', in J. J. C. Smart and B. Williams (eds.), *Utilitarianism: For and Against*. Cambridge University Press, Cambridge.

SMITH, M. (1988). 'Consequentialism and Moral Character' (unpublished manuscript, Philosophy Dept., Monash University, delivered to the philosophical society, Oxford University, June 1988).

SMITH, R. (1985). *Liberalism and American Constitutional Law*. Harvard University Press, Cambridge, Mass.

SOMMERS, C. (1987). 'Filial Morality', in Kittay and Meyers (1987).

STEINER, H. (1977). 'The Natural Right to the Means of Production', *Philosophical Quarterly*, 27/106: 41–9.

—— (1981). 'Liberty and Equality', *Political Studies*, 29/4: 555–69.

—— (1983). 'How Free: Computing Personal Liberty', in A. P. Griffiths (ed.), *On Liberty*. Cambridge University Press, Cambridge.

STERBA, J. (1988). *How to Make People Just: A Practical Reconciliation of Alternative Conceptions of Justice*. Rowman and Littlefield, Totowa, NJ.

STIEHM, J. (1983). 'The Unit of Political Analysis: Our Aristotelian Hangover', in S. Harding and M. Hintikka (eds.), *Discovering Reality*. Reidel, Dordrecht.

STOCKER, M. (1987). 'Duty and Friendship: Towards a Synthesis of Gilligan's Contrastive Moral Concepts', in Kittay and Meyers (1987).

STOUT, J. (1986). 'Liberal Society and the Languages of Morals', *Soundings*, 69/1–2: 32–59.

SULLIVAN, W. (1982). *Reconstructing Public Philosophy*. University of California Press, Berkeley, Calif.

SUMNER, L. W. (1987). *The Moral Foundation of Rights*. Oxford University Press, Oxford.

TAUB, N., and SCHNEIDER, E. (1982). 'Perspectives on Women's Subordination and the Role of Law', in D. Kairys (ed.), *The Politics of Law*. Pantheon, New York.

TAYLOR, C. (1979). *Hegel and Modern Society*. Cambridge University Press, Cambridge.

TAYLOR, C. (1985). *Philosophy and the Human Sciences: Philosophical Papers*, ii. Cambridge University Press, Cambridge.

—— (1986). 'Alternative Futures: Legitimacy, Identity and Alienation in Late Twentieth Century Canada', in A. Cairns and C. Williams (eds.), *Constitutionalism, Citizenship and Society in Canada*. University of Toronto Press, Toronto.

—— (1989). 'Cross-Purposes: The Liberal–Communitarian Debate', in N. Rosenblum (ed.), *Liberalism and the Moral Life*. Harvard University Press, Cambridge, Mass.

TONG, R. (1989). *Feminist Thought: A Comprehensive Introduction*. Westview Press, Boulder, Colo.

TRONTO, J. (1987). 'Beyond Gender Difference to a Theory of Care', *Signs*, 12/4: 644–63.

UNGER, R. (1984). *Knowledge and Politics*. Macmillan, New York.

VAN DER VEEN, R., and VAN PARIJS, P. (1985). 'Entitlement Theories of Justice', *Economics and Philosophy*, 1/1: 69–81.

VAN DYKE, V. (1975). 'Justice as Fairness: For Groups?', *American Political Science Review*, 69: 607–14.

VARIAN, H. (1985). 'Dworkin on Equality of Resources', *Economics and Philosophy*, 1/1: 110–25.

VOGEL, U. (1988). 'When the Earth Belonged to All: The Land Question in Eighteenth Century Justifications of Private Property', *Political Studies*, 36/1: 102–22.

WALDRON, J. (1986). 'Welfare and the Images of Charity', *Philosophical Quarterly*, 36/145: 463–82.

—— (1987). 'Theoretical Foundations of Liberalism', *Philosophical Quarterly*, 37/147: 127–50.

—— (1989). 'Autonomy and Perfectionism in Raz's *Morality of Freedom*', *Southern California Law Review*, 62/3–4: 1097–152.

WALZER, M. (1983). *Spheres of Justice: A Defence of Pluralism and Equality*. Blackwell, Oxford.

—— (1990). 'The Communitarian Critique of Liberalism', *Political Theory*, 18/1: 6–23.

WEALE, A. (1982). *Political Theory and Social Policy*. Macmillan, London.

WEITZMAN, L. (1985). *The Divorce Revolution: The Unexpected Social and Economic Consequences for Women and Children in America*. Free Press, New York.

WENDELL, S. (1987). 'A (Qualified) Defense of Liberal Feminism', *Hypatia*, 2/2: 65–93.

WILLIAMS, B. (1971). 'The Idea of Equality', in H. Bedau (ed.), *Justice and Equality*. Prentice-Hall, Englewood Cliffs, NJ.

—— (1972). *Morality: An Introduction to Ethics*. Harper and Row, New York.

—— (1973). 'A Critique of Utilitarianism', in J. J. C. Smart and B.

Williams (eds.), *Utilitarianism: For and Against*. Cambridge University Press, Cambridge.

—— (1981). *Moral Luck*. Cambridge University Press, Cambridge.

—— (1985). *Ethics and the Limits of Philosophy*. Fontana Press, London.

WILSON, L. (1988). 'Is a "Feminine" Ethic Enough?', *Atlantis*, 13/2: 15–23.

WOLFF, R. (1977). *Understanding Rawls*. Princeton University Press, Princeton, NJ.

WOLGAST, E. (1987). *The Grammar of Justice*. Cornell University Press, Ithaca, NY.

WOLIN, S. (1960). *Politics and Vision*. Little Brown, Boston, Mass.

WOOD, A. (1972). 'The Marxian Critique of Justice', *Philosophy and Public Affairs*, 1/3: 244–82.

—— (1979). 'Marx on Right and Justice', *Philosophy and Public Affairs*, 8/3: 267–95.

—— (1981). 'Marx and Equality', in J. Mepham and D. H. Ruben (eds.), *Issues in Marxist Philosophy*, iv. Harvester Press, Brighton.

—— (1984). 'Justice and Class Interests', *Philosophica*, 33/1: 9–32.

YOUNG, I. (1981). 'Toward a Critical Theory of Justice', *Social Theory and Practice*, 7/3: 279–302.

—— (1987). 'Impartiality and the Civic Public', in S. Benhabib and D. Cornell (eds.), *Feminism as Critique*. University of Minnesota Press, Minneapolis, Minn.

—— (1989). 'Polity and Group Difference: A Critique of the Ideal of Universal Citizenship', *Ethics*, 99/2: 250–74.

ZARETSKY, E. (1982). 'The Place of the Family in the Origins of the Welfare State', in B. Thorne and M. Yalom (eds.), *Rethinking the Family: Some Feminist Questions*. Longman, New York.

Index